Contents

Introduction	1
1 Childhood and early life	11
2 Success	24
3 Glasgow	45
4 Aberdeen and the First World War	64
5 Oxford and 'big science'	84
6 Monetary and other theories	108
7 New Britain	133
8 The darkest days	156
9 The post-war world	178
Epilogue: The Frederick Soddy Trust	197
Bibliography	207
Index	221

Acknowledgements

I would like to thank the Frederick Soddy Trustees who commissioned this book and who supported me, as the Frederick Soddy Research Fellow, during its writing. I would also like to thank the staff and faculty of the School of Cultural and Community Studies, University of Sussex, my institutional 'home', and especially Stuart Laing, the Dean, for much of the work. Like all Sussex workers I am very grateful for the support and patience of the library staff, particularly Bett Inglis. I owe a special debt to the hospitality of the New Atlantis Group and Ditchling and to their close associate, the late Reginald Wrugh.

A large number of libraries and collections made material available to me and treated me with unfailing kindness. My thanks to the librarians, archivists and staff of: Aberdeen University Library, Department of Special Collections; the Bodleian Library, Oxford; Cambridge University Library; East Sussex Record Office; the Fawcett Library, London; McGill University, Montreal; Mitchell Library, Glasgow; the Museum of the History of Science, Oxford; New Atlantis, Ditchling; the Royal Institution; University College, London; the University of Bristol; the University of Glasgow Archives; the University of Illinois Library at Urbana-Champaign (Rare Books Room); the University of Keele Library; the University of Strathclyde; the University of Sussex. I would also like to thank the Royal Society. Materials from their archives are reproduced by kind permission of the President and Council of the Royal Society of London.

Finally, I would like to thank the editorial staff of OUP for their support in publishing this book; Peter Dickens, School of Social Sciences, University of Sussex, for endless intellectual stimulation; the Scratch Club at the University of Edinburgh for encouragement at a very early stage, and Alun Howkins for his unfailing belief in me and Soddy.

the Cairngorms with a note asking me to read it and return it before we went and he wished me fair weather. Clearly somewhere in that complex personality there was a warm friendly side to his nature which he very rarely showed in Oxford and of which I had just had a glimpse.

I never saw him again; nor did I meet any of his intimates—if such there were—but these personal memories have remained with me ever since and from time to time I've seen references to him, usually in relation to some statement he had uttered or written generally, widely at odds with the received wisdom of the day. And, of course, I was aware of his association with Le Play Society. So I was greatly surprised to be asked to write a Foreword to this book. It was an invitation I could not refuse subject to the proviso that I could see it in typescript before I decided, because I was not prepared to assume this role if it did not seem to me to be objective, aiming neither to praise nor to condemn. Anything less than this would, in my view, be a betrayal of all that Soddy stood for. Having read the book but none of the references, I gain the feeling that Dr Merricks has done full justice to her subject and the times in which he lived.

But that is not all. In giving us this life of Soddy she has given a credible picture of the United Kingdom in the early years of this century, in a period when science made great advances and intruded increasingly into the life of individuals and the community, of the problems of the funding of science, of the painfully slow reformation of Oxford University, of political, social, and economic change in the country—all as seen through the eyes of, or directly experienced by, this unusual man to whom Terence's words '*Homo sum, nil a me alienum puto*' are truly applicable. I commend this book to the reader.

Oxford
September 1995

Dainton of Hallam Moors
(Formerly Dr Lee's Professor of Chemistry
1970–1973)

My acute disappointment can be imagined when a few days after arrival at Oxford I was summoned by my chemistry tutor to discuss my programme of work and, referring to lectures, he strongly advised me *not* to go to those given by the then head of his department, the Old Chemistry Department, then Dr Lee's Professor of Chemistry, one Frederick Soddy. I shall always be glad I disregarded my tutor's advice. This was not solely due to seeing and hearing at first hand some of the extraordinary work Soddy had done, using the most primitive detectors (I well remember his dexterous handling of a gold leaf electroscope), together with the sense of actually being present when the work was done and sharing in the successes and failures and so beginning to learn something of the joy of discovery. There were also Soddy's face and gait; the former being finely chiselled with very steady eyes betokening a man of high principles unwilling to compromise truth as he saw it at whatever cost to personal relations; his gait and bearing suggesting also a very fit man, confident of his own powers and afraid of no man.

I soon found that the words used about him in Oxford were typically 'obstinate and uncooperative', 'doesn't do research any more', 'a crackpot, wasting his time with solid geometry problems and deflecting much needed workshop activity from its proper task', 'is absorbed by the dotty idea that national economics can be analysed on thermodynamic principles', 'espouses fringe political movements, like the Canadian Major Douglas' Social Credit Party, and New Britain', and finally 'a disappointment to himself as well as to us'. Not a good word was said about him despite his having redesigned the antiquated laboratories and their refurbishment down to the smallest detail, and thereby benefiting those who were often his greatest critics as well as the students.

All this had nothing to do with undergraduates and Soddy was the subject of few, if any, of our conservations. But one day when I was doing practical qualitative analysis he strode characteristically and purposefully down the central isle of the teaching laboratory looking neither to left nor right. I expected him to pass my peninsular bench but instead he stopped, turned towards me and said 'Where are you going with that thing?' The 'thing' to which he pointed was a newly acquired Bergan rucksack intended for use in the summer of 1934 when I and two colleagues planned to spend three weeks in the Cairngorms. The mention of these hills brought a smile to Soddy's face and we talked about them, about Deeside and Donside, and Aberdeen itself during which he reminisced with evident pleasure. He asked how I would face the high cost of travel from Sheffield by train to Kirriemuir from which we proposed to begin our walk. I said I thought that I might get an A. C. Irvine Travel Fund Award which produced another smile and then he departed. Two days later he left a brown envelope containing his copy of Seton Gordon's book about

Foreword
The Lord Dainton FRS

1932 has often been described as the *annus mirabilis* of the Cavendish Laboratory of Cambridge University. The well known photograph of the staff and research students taken that year is ample proof of this because, amongst those present, were the following persons who either had been or would be awarded the Nobel Prize for Physics; (Sir) J. J. Thomson, (Lord) Rutherford, J. D. Cockcroft, C. T. R. Wilson, E. T. S. Walton, P. M. S. Blackett, James Chadwick, P. A. M. Dirac, Neville Mott, F. W. Aston, and E. O. Lawrence. The primary preoccupation of all these men and their colleagues was to understand the nature of the nuclei of all known elements, why some nuclei are quite stable whilst others decompose giving out either packets of energy in the form of radiation (γ-rays), positively charged helium nuclei (α-rays), or negatively charged electrons (β-rays). To this day this remains the quest of theoretical and experimental physicists the world over and for this purpose they often build and use great particle accelerating machines, operating at energies inaccessible, indeed, not even dreamt of sixty years ago.

It was also an *annus mirabilis* for me. In the first place whilst still a school boy I heard James Chadwick, in a lecture at the University of Sheffield, describe this discovery of the neutron and then speculate on its role together with that of protons as nuclear constituents. I was enthralled by the experience. Secondly, in December I was offered an award by St John's College, Oxford which, provided I could augment it from other sources, would enable me to achieve my ambition of going to Oxford to read chemistry. In the following nine months, during which my overriding objective was to teach myself enough Latin to be admitted to the University of Oxford, my approach to science was more relaxed and I read widely in the Sheffield Public Reference Library. There, amongst many other treasures, I came across various books on radioactivity including that by Soddy *The Interpretation of the Atom* (1932). I began to appreciate what a seminal role his experiments made in Canada and Scotland had played in enlarging our knowledge of the atomic nucleus. They had led him to describe the all important concept of the 'Law of radioactive displacement', and the associated concept of isotopes i.e. elements of the same positive nuclear charge but different mass which therefore were chemically identical. And so I came to realize what a great man Soddy was and that his place in science is assured for ever.

Oxford University Press, Walton Street, Oxford OX2 6DP
Oxford   New York
Athens   Auckland   Bangkok   Bombay
Calcutta   Cape Town   Dar es Salaam   Delhi
Florence   Hong Kong   Istanbul   Karachi
Kuala Lumpur   Madras   Madrid   Melbourne
Mexico City   Nairobi   Paris   Singapore
Taipei   Tokyo   Toronto
and associated companies in
Berlin   Ibadan

Oxford is a trade mark of Oxford University Press

Published in the United States
by Oxford University Press Inc., New York

© Linda Merricks, 1996

All rights reserved. No part of this publication may be
reproduced, stored in a retrieval system, or transmitted, in any
form or by any means, without the prior permission in writing of Oxford
University Press. Within the UK, exceptions are allowed in respect of any
fair dealing for the purpose of research or private study, or criticism or
review, as permitted under the Copyright, Designs and Patents Act, 1988, or
in the case of reprographic reproduction in accordance with the terms of
licences issued by the Copyright Licensing Agency. Enquiries concerning
reproduction outside those terms and in other countries should be sent to
the Rights Department, Oxford University Press, at the address above.

This book is sold subject to the condition that it shall not,
by way of trade or otherwise, be lent, re-sold, hired out, or otherwise
circulated without the publisher's prior consent in any form of binding
or cover other than that in which it is published and without a similar
condition including this condition being imposed
on the subsequent purchaser.

A catalogue record for this book is available from the British Library

Library of Congress Cataloging in Publication Data
Merricks, Linda.
The world made new : Frederick Soddy, science, politics, and
environment / Linda Merricks.
1. Soddy, Frederick, 1877–1956.   2. Radioactivity–History.
3. Radiochemistry–History.   4. Physicists–Great Britain–
Biography.   I. Title.
QC16.S75M47 1996   540'.92–dc20 [B]   95–42047

ISBN 0 19 855934 8

Typeset by Palimpsest Book Production Limited,
Polmont, Stirlingshire
Printed in Great Britain by
Bookcraft Bath Ltd
Midsomer Norton, Avon

The World Made New
Frederick Soddy, Science, Politics, and Environment

LINDA MERRICKS

Formerly Frederick Soddy Research Fellow, University of Sussex, and Lecturer in History, School of English and American Studies, University of Sussex

Oxford New York Tokyo
OXFORD UNIVERSITY PRESS
1996

For Alun

Introduction

Biography is a somewhat peculiar branch of history. History, conventionally, looks at and tries to explain a particular community or culture, whether a parliament or food riots, through a particular period of time. In this it is selective, choosing what is relevant to that specific organisation and the relationships of individuals to it, and history therefore tends to remain within a set of parameters defined by the topic. For example, historians of parliament rarely look at the popular culture of their period. In contrast to this, the only defining feature of a biography is that it centres on one person. This can lead to the most unexpected areas of investigation since any individual will intersect with a number of different societies in the course of a lifetime. What are fascinating are the connections between these different areas and the ways in which they interact with each other and with the subject of the biography.

This is especially true of the subject of this work. Frederick Soddy was born in 1877 and so was one of the very early generation of English atomic scientists. He was different from most of his scientific contemporaries because as an undergraduate in 1896 he went to Oxford, rather than Cambridge which even at that early stage had established a reputation for scientific excellence. He was a chemist, not a physicist, which was to lead to difficulties in the development of his career. He was, above all, an individualist. At the beginning of his career this was an advantage. It gave him the necessary impetus to travel to Canada in search of what he regarded as a suitable post, a move which led to his association with Ernest Rutherford. It also made him temperamentally unsuited to the 'big science' which began to develop between the wars, with its need for teamwork and co-operation. Because of the notion that has developed during the twentieth century that, with a few exceptions, the natural sciences are above (or beyond) political involvement, it now seems paradoxical that he was also a socialist for many years with a developed sense of the responsibility of the scientist. A result of this was that, before the Great War, he was concerned about the potential effects of the release of atomic energy.

This early period of his life is fairly well recorded. He worked with Rutherford at McGill in Montreal from 1901–1903, during which short period they published the eight papers which set out the 'Disintegration Theory of Atomic Transmutation', the first work which showed experimentally that the atom was not indivisible and therefore the work on which subsequent discoveries in atomic science were based.

Rutherford received the Nobel Prize for this work, but Soddy had to wait for his. On his return from Canada he spent a short time working with William Ramsay in London and identified the particles expelled from radium as helium. In 1904, he then moved to Glasgow for a brief but successful period up to 1914. Here he worked on the displacements in the periodic table brought about by radioactive changes and published his theory of chemically identical elements with different atomic weights which he called isotopes. For this work he was awarded the Nobel Prize for 1921.

During this period his personal life was as successful as his professional life. He married Winifred Beilby, with whom he was to spend nearly thirty very happy years. He developed a love of climbing and walking in mountains which remained with him for the whole of his life. He was also introduced to politics and began to develop ideas about the relationship of science and society. These ideas have since become commonly known as the 'social responsibility of science', but were considered novel and somewhat odd at the time.

In 1914, just before the outbreak of war, he moved to Aberdeen. This signalled the end of his 'honeymoon period' in science. He was beset by difficulties in dealing with military and scientific authorities throughout the war which, together with the demands on laboratory space for research into the development of the necessities of warfare, interfered with any other projects. He also, like many others, began to consider the ethical problems of discoveries which could bring untold wealth and happiness, but could also destroy the world. As Fritz Haber says in Tony Harrison's verse drama *Square Rounds*,[1]

> I'm only the inventor how can I guarantee
> no one will turn my nitrates into TNT?
> Duality reigns. It wasn't my decision
> to have my ammonia turned into ammunition.

Haber had discovered the nitrogen fixation process, which was the basis of artificial fertilisers, but which could also be used for explosives. (He was also to develop poison gas, for which there was no such justification, but believed it would hasten the end of the war.) Soddy's difficulty was the potential of atomic energy which could be even worse than TNT.

From the Montreal period Soddy and Rutherford had attempted to calculate the amount of energy released in radioactive transformations and had come to quite terrifying answers. Soddy explained these quantities to a popular audience in 1909 when he wrote that the amount of energy given out by radium in a year was 'a quarter of a million times as much as is given out in the combustion of an equal amount of coal.'[2] If this power could be harnessed, poverty could be eradicated and the Garden of Eden reinstated on earth. Both believed that the actual harnessing of this power was likely to remain impossible. However, Rutherford insisted that if any use was made of the discovery it would be none of his business—like Haber, he was 'only the inventor'. Soddy felt very differently and began to fear that, under the pressures of wartime, atomic energy might be developed and used for weapons. Soddy's fears were picked up by H. G. Wells and put into novel form in *The World Set Free*, but otherwise had little effect, except on his own development.

He began to consider the world he inhabited which, he believed, was too irresponsible for increasingly powerful scientific discoveries, and from here he began to wonder how improvement was possible. After the war, in 1919, he was appointed to the Dr Lee's Chair of Chemistry at Oxford University, with the expectation that he would found a research school of radiochemistry where he would continue to consider this problem. At the beginning of the 1920s he had contact with a number of groups and organisations including the Sociological Society, the National Union of Scientific Workers, the Labour Party, National Guilds and the Social Credit Movement, but none seemed to him to provide a solution. His interests increasingly converged around the economic problems of the period which he came to believe were behind the social inequalities.

He came to believe, with Major Douglas and Arthur Kitson, that currency reform was the first essential and as a result he published a number of works on his theories of money, its circulation and the role of banks. These works never brought the attention he felt they deserved, but led him into the strange and exciting world of inter-war London with its fluid and interlocking groupings, its celebrities and nonentities, all at first centred on A. R. Orage and the *New Age*, and then developing into a number of different groups and organisations with different aims. This led to a close association with the now little known New Europe, New Britain and other groups who shared the common aim of a new, peaceful society through a combination of monetary reform and a European federation. His affiliation to these groups lasted throughout the inter-war period and until the 1950s.

His scientific output was disappointing during this period. His Oxford appointment had been made in the expectation that he would set up a school of radioactivity to rival that of Rutherford at Cambridge. However, there were two substantial obstacles which coalesced to prevent any such development. Firstly, if not most importantly, his socio-political concerns were given a great deal of his time. Secondly, and perhaps this was the cause or effect of the first obstacle, he was too much of an individualist to conform to the byzantine necessities of the bureaucracy of an ancient self-governing university. The characteristics which had set him off on his career with a trip to Montreal were totally unsuited to Oxford in the 1920s and 1930s. He found it impossible to negotiate for resources of space or money. He had trouble with staff and barely saw a student. It took him nearly ten years even to begin to equip a laboratory suitable for research. Thirdly, it has to be said that Oxford found him an uncomfortable bedfellow and responded with a policy of gentlemanly stonewalling, an all too familiar occurrence in the history of English institutions. For example, a younger colleague remembered that 'during the 1930s plans were being laid for a University Physical Chemistry Laboratory . . . However, it was generally agreed that it would be disastrous if Soddy gained control . . . and nothing was done until he retired.'[3]

After the Second World War, the bombing of Hiroshima and Nagasaki led to renewed anxieties about the effects of the power he had fathered. His position as the last remaining English atomic scientist of his generation caused his short-lived fame which led to a radio broadcast and the publication of his *Memoirs*. At this time he also revived his sociological interests through his contact with the Le Play Society. This had originated in Patrick Geddes' Sociological Society but the evolution of the disciplines of sociology and geography through the 1930s and afterwards had reduced the necessity for such amateur approaches. The Society had tended to stagnate and by the post-Second World War period, when Soddy's involvement was closest, the membership was small and close-knit, and ageing. In this company, Soddy was to find pleasure and companionship during his last years.

Soddy always felt that he did not receive the recognition he deserved. While this complaint is not uncommon, in his case there seems to be some justification. Awareness of his works falls almost entirely into two periods. From his time as an undergraduate until the early 1920s, Soddy attracted the attention expected by any young scientist, even if he was, initially, considered as Rutherford's junior partner. His articles appeared regularly in *Nature* and other journals and attracted attention even (as mentioned above) by the non-specialist H. G. Wells. Soddy's reputation would have been enhanced perhaps by the publication of an academic account of his researches instead of the popular *Interpretation of Radium*, but, as the award of the Nobel Prize showed, his reputation was secure.

It is with his change of direction in 1920 that difficulties can be seen to begin and he disappears from the record, descending into obscurity. His turning from science elicited little, if any, sympathy. This was, after all, the beginning of an heroic period in science when the needs of research into radioactivity led to demands for hitherto unknown sums of money, and consequently to a public awareness of the importance of the subject. Instead of being involved in this, Soddy chose to publish lectures, books and articles in an area which itself was struggling in the throes of development into a recognisable discipline—the creation of a theoretically sound science of economics. Here he was regarded as an amateur and an interloper, unaware of the mysteries of the new profession. As a result, his works were rarely reviewed and, even when noticed, tended to be included amongst the work of currency cranks. In this he was not alone: other members of the same groups and organisations felt the same way.

As long as he was in the company of others receiving similar treatment, Soddy seems to have complained but not felt that it was especially unusual. 'To him that hath shall much be given' is the byword of publicity of all kinds. So long as one is successful and, in the sciences, making exciting discoveries, attention in both professional and popular works will seem normal. The unsuccessful or the ordinary will be ignored. This is to over-simplify and ignores the effect of this attention on the recipient; the power of the media to create heroes is now well known. Possibly, if the monetary theories had been considered more interesting than the theories emanating from the newly established London School of Economics, history would have recognised them. However, public attention then and history now have given their judgement.

It is during his second period of public notice that Soddy seems to have been particularly anxious about what he came to describe as a conspiracy of silence. His appearances after the Second World War, when he referred to his fears about atomic energy, resulted in some publicity, although nearly all in less central publications. However, by this time he had come to the notice of Muriel Howorth, an amateur enthusiast for atomic energy and founder of 'The Institute of Atomic Energy for the Layman'. She had decided to produce his *Memoirs* and then to write his biography which, against his continuing residual resistance, she did, completing *Pioneer Research on the Atom* in 1958, two years after his death. These works share the laudable aim of restoring Soddy to public notice but are marred by their partisan approach. Howorth's position is quite straightforward. Rutherford in some way managed to attract all attention and credit for the discoveries made jointly with Soddy, and as a result his career flourished while Soddy's declined. Whatever the truth of this assertion, such an approach could only harm Soddy's reputation, as he himself realised.

Since his death, Soddy has received more widespread notice. As is to be expected, his obituaries were used as an occasion for assessment. These were divided in a way which has influenced writing on him ever since. With rare exceptions, they began with a firm recognition of his achievements in science. His obituary in *The Times*, for example, talked of the 'classic series of experiments on radioactivity' with Rutherford at

McGill and went on to praise his work with Ramsay. *The Interpretation of the Atom* was described as 'one of the seminal works on the subject in its period'.[4] A full tribute to and assessment of his work was given by Professor F. A. Paneth in *Nature*. Paneth stressed, as had *The Times*, Soddy's great scientific achievement. He was 'a brilliant intellect, an experimenter second to none among the founders of radio-chemistry, and an uncompromising champion of his ideals'.[5] However, Paneth also touched on the other aspect of Soddy's life—the virtual end of his scientific career after the Great War. To Paneth this was a tragic but understandable result of Soddy's long-term commitment to social reform dating from his advocacy of Irish home rule and votes for women before the Great War. However, others were less generous. A particularly vitriolic 'tribute' appeared in *The Tablet* under the title 'The Aberration of Genius' by Colin Clarke.[6] Here, although his early scientific work is given fulsome praise, Soddy's later career is subject to an attack which is at best intemperate and at worse downright insulting. Soddy's lectures at Oxford were 'almost valueless, disfigured by extraordinary and unjustified speculations, whose baselessness was apparent even to undergraduates.' His social and political beliefs get even shorter shift from Clarke who, perhaps rather strangely in view of his insistence on the 'baselessness' of Soddy's science, seems to rely quite heavily on papal pronouncements. This view leads to the conclusion that Soddy was the first example of the 'distinguished scientist who is getting a bit tired of his science, and who seeks to give the world the benefit of his views on social questions . . .'

This view of Soddy's career as divided into two different and hostile halves has continued to shape the writing on him. In addition, the power with which he held his later views has overshadowed and obscured his real achievements in science. As Paneth wrote in his tribute in *Nature*,[7] 'Anybody who studies the original papers will easily recognize the decisive part [Soddy] must have played in this joint work. . . . Nevertheless in later books Soddy's name sometimes no longer appears in this context: to Rutherford is attributed the sole merit.' It was these 'omissions' which lead to Howorth's biography, which we have already briefly touched upon, and which, because of its tone, did little to alter Soddy's image. As Paneth writes, 'It is most regrettable that these unfounded grievances of an old and disgruntled Soddy have found their way into recent books.'[8]

His reputation as a scientist has been given support in recent years by the account of the 'prehistory' of atomic energy given as an introduction by Kenneth Jay to Margaret Gowing's standard work, *Britain and Atomic Energy, 1939–1945*.[9] In this, Soddy is given the attention he deserves both as a partner of Rutherford and for his work on isotopes. Then, in 1977, there was a special 'Soddy Session' at the Fifteenth International Congress of the History of Science in Edinburgh, chaired by Thaddeus J. Trenn, to celebrate the centenary of Soddy's birth. The papers presented at this session were published in *The British Journal for the History of Science* in 1979 and were then reprinted, with further essays, in *Frederick Soddy (1877–1956): Early Pioneer in Radiochemistry*, edited by George B. Kauffman in 1986.[10] These essays give detailed accounts of Soddy's scientific work, largely redressing the imbalance of Howorth's accounts. Most recently, Professor Mansel Davies, in 'Frederick Soddy: The Scientist as Prophet' in *Annals of Science* (1992) has written about Soddy as a scientist and given some indication of the wider implications of his works.[11]

This leads to another area in which Soddy's influence has recently been discovered. This is within the developing subject of pacifist or Green history. Brian Easlea, in *Fathering the Unthinkable*, has shown that while Rutherford may be described as the 'father' of atomic theory, Soddy was at the start of a different tradition, now known as 'the social responsibility of science'.[12] In a less directly critical account, I. F. Clark in *Voices Prophesying War* has again used the story of Soddy's influence over H. G. Wells, here maintaining that Soddy is actually portrayed in the novel as the Professor in Edinburgh who 'makes one of the most prophetic statements in the history of science fiction: "If at a word, in one instant, I could suddenly release that energy [of the uranium he is demonstrating in his lecture] here and now, it would blow us and everything about us to fragments".' In Clark's account Soddy is actually given more space than Rutherford, a sign of his increasing importance in a world faced with the real prospect of the release of such energy.[13]

Other recent writers in this area have emphasised that Soddy's prophetic insights into the potential of atomic weapons came not only from his science but were also 'supplemented by emotional involvement, intense creativity, and social awareness ... assisted by sources of knowledge entirely outside the usual domain of science', ranging from 'contemporary politics and science fiction to religion, mythology, and ancient alchemy'.[14] Selove argues further that Soddy's interest in alchemy enabled him to predict transmutation before the actual discovery, but the evidence for this claim is uncertain. What is perhaps clearer is that Soddy used his wide reading to construct a theory of cyclical history—that the world had already existed—a notion picked up by Spencer R. Weart in *Nuclear Fear*. Both these writers refer to the unusual side of Soddy's early reading and lectures, his interest in the very earliest history of anything that could be called science.[15] Thus, in McGill he had lectured on alchemy while working with Rutherford.

Soddy's work has been placed within another kind of tradition by Anna Bramwell in *Ecology in the 20th Century: A History*.[16] Here she describes Soddy as an 'economic ecologist', one of those who 'call to conserve scarce resources' in a closed universe. Bramwell argues that thinkers in this field include the anarchist Kropotkin and the sociologist Patrick Geddes, both of whom, like Soddy, based many of their ideas on Ruskin. The basis on which Soddy can be defined as an ecologist is, for Bramwell, 'the way he perceived agriculture as the "key industry", and the cycle sun–soil–food as "the internal energy of life".'[17] Bramwell also looks at Soddy's economic theories, using an approach given a more detailed treatment in Juan Martinez-Alier's *Ecological Economics*.[18] Here, Soddy's economic theory is given a detailed and careful consideration which emphasises the extent to which he was already suggesting ideas about wealth as a flow, not a store, and about the role of energy, ideas which are only recently being applied to other questions.[19]

All these works have been enormously useful for the present book. However, there remain two gaps, which I hope are dealt with here, at least as far as is now possible. The first is in relating all these different aspects to each other, within the context of the society in which they were happening, and to the man himself. This has been the purpose of the present work. Some of what follows relies on the works cited here and on others, especially Soddy's own writings. However, the importance of this work lies in the connections made here between these various facets of a personality, in providing an explanation for Soddy's shift from science through politics and sociology into monetary theory, and finally into the Le Play Society. Soddy's importance rests only in part on what he actually did; the other reason is the sheer diversity of his interests throughout his life and the way in which he gave then some coherence. These interests also illustrate the times through which he lived. He was in many ways typical of his generation in each stage of his life, although sometimes rather ahead of his times. He entered science at the end of the nineteenth century, just as the subject was gaining academic and more general recognition and so beginning to develop. He then expressed fears about the consequences of fighting, during a war in which many of the intellectuals were doing the same, even if they nearly all settled for 'their duty'. After the war, like Maynard Keynes, G. D. H. Cole, much of the government and many others, he turned his attention to the tragedy of unemployment, the slump and therefore to the economy. In the early twenties he was involved in left wing politics and the early days of his union, the National Union of Scientific Workers. He spent time with a self styled eastern European mystic when people like A. R. Orage were finding their own gurus in Paris or elsewhere, and was on the fringes of the Fascist movement in the early thirties, looking for a peaceful Europe but fearing another war. He even travelled to the East in the late thirties, pre-dating the hippies by nearly thirty years, in a search for solace after the death of his wife. After the Second World War and the awfulness of the atomic bomb, he was loud in his condemnation of this new form of warfare, stressing the dangers it posed to the whole world.

Unfortunately, what is missing from this book is any detailed account of his personal life. The reasons for this are quite simple. Although I have met a number of people who knew him, they are by definition not only themselves getting older, but also those who met him late in his life. They have given some indication of the kind of man he was, but the necessary records for a real sense of his likes and dislikes, his emotional involvements, his recreational reading, even his friends, are missing. It is probable that some never existed; he never appears as the kind of person who kept a detailed diary, for example. But even if he had, he himself censored all his papers in the years leading up to his death. He removed anything of a personal nature and then sorted and ordered what remained. Even these papers, however, do not all still exist. He left them to his literary executrix, Muriel Howorth, who had a somewhat eccentric filing system, consisting of cardboard boxes under her sink. The inevitable result of this is that, accidentally or deliberately, some of the remaining papers were lost before the remnants were deposited in the Bodleian Library. Most of these relate to his scientific career. There are very few which shed any light on his attitudes and emotions, except again, where science was the subject.

Luckily, others were more careful. Soddy was a prolific and methodical letter writer. His correspondence with Rutherford spanned twenty years between 1902 and 1922, and these letters are held in the Cambridge University Archive. His association with the New Europe Group between 1931 and 1954 resulted in a large collection of letters and lectures and other materials. The period in between is the time when he was developing his monetary theories and published a mass of material through which his intellectual development, if not his emotions, can be traced, and the survival of his notebooks and lecture notes provides a body of illuminating material. However, for his character, his relationships with his wife, his religious and political beliefs, it is necessary to judge him by his actions, his published writings, and the opinions of others.

Notes and references

1. Harrison, T. (1992). *Square Rounds*, p.28. Faber and Faber. (This drama was produced in 1992 at the National Theatre in London.)
2. Soddy, F. (1909). The Energy of Radium. *Harper's Monthly*, **120**, December 1909, p. 54.
3. Letter from R. P. Bell to the author, dated 3 February 1991.
4. *The Times*, 24 September 1956.

5. *Nature*, 23 November 1957, **180**, 1087.
6. Clarke, C. (1956). The Aberration of Genius. The Work of Frederick Soddy. *The Tablet*, 13 October 1956, pp. 298–300.
7. *Nature*, 23 November 1957, **180**, 1085.
8. *Nature*, 23 November 1957, **180**, 1086.
9. Gowing, M. (1964). *Britain and Atomic Energy, 1939–1945*, pp. 3–32. Macmillan.
10. *The British Journal for the History of Science*, November 1979, **12**, 3, No. 42, 245–88; Kauffman, G. B. (1986). *Frederick Soddy (1877–1956): Early Pioneer in Radiochemistry*. Reidel, Dordrecht.
11. Davies, M. (1992). Frederick Soddy: The Scientist as Prophet. *Annals of Science*, **49**, 351–67.
12. Easlea, B. (1983). *Fathering the Unthinkable: Masculinity, Scientists and the Nuclear Arms Race*, pp. 61–2 and Chapter 2 *passim*. Pluto Press.
13. Clark, I. F. (1992). *Voices Prophesying War: Future Wars 1763–3749*, p. 157. Oxford University Press.
14. Selove, R. E. (1989). From Alchemy to Atomic War: Frederick Soddy's 'Technological Assessment' of Atomic Energy, 1900–1915. *Science, Technology and Human Values*, **14**, No. 2, 163–94, especially 164.
15. This is emphasised by Weart. See Weart, S. R. (1985). The Heyday of Myth and Cliche. *Bulletin of Atomic Scientists*, **41**, 38–43, especially 39.
16. Bramwell, A. (1989). *Ecology in the 20th Century: A history*, p.82–5, but also *passim*. Yale University Press, New Haven.
17. Bramwell, A. (1989). *Ecology in the 20th Century: A history*, p. 84. Yale University Press, New Haven.
18. Martinez-Alier, J. with Schlupmann, K. (1987). *Ecological Economics: Energy, Environment and Society*. Basil Blackwell, Oxford.
19. Martinez-Alier, J. with Schlupmann, K. (1987). *Ecological Economics: Energy, Environment and Society*, Chapter 9. Basil Blackwell, Oxford.

CHAPTER 1
Childhood and early life

There was nothing in his family history to suggest that Frederick Soddy would become a scientist. None of his ancestors had managed to get as far as university, and there was certainly nothing at all scientific in his background. The Soddy family had originated in the Bunhills district of London and family tradition has it that they were of Huguenot extraction, the family having fled to London in the seventeenth century.[1] Frederick's father, Benjamin Soddy, had moved his family from Walworth in South London to Eastbourne in 1875. No reason has been recorded for this move, but the reputation of the south coast for its health-giving properties prompts the speculation that illness was the likely cause. Although there is no direct evidence, either Benjamin or Hannah, his second wife, might have been the victim. At the time of the move, Benjamin was only in his early fifties, but he retired from full-time employment as a corn merchant in favour of a three-day-a-week position at the London Corn Exchange as a buyer for 'Messrs. Tilling, The London Omnibus Company'.[2] However, he lived until 1911 and so perhaps this semi-retirement was a matter of personal choice. On the other hand, Hannah was only in her early thirties and had three young sons, John, Joseph and Thomas, all under three years of age when they moved, and then Frederick was born in 1877. Again, there is no mention of disease but such frequent pregnancies apparently took their toll of Hannah's health and she died when Frederick, the youngest child, was only eighteen months old.

Having decided to move south, Eastbourne was an obvious choice for the Soddy family. Like other south coast towns it had a reputation for a healthy environment, but this was not enough to ensure its development which had almost entirely depended on the railway link with London. The regular service made it attractive to men like Soddy, needing to commute regularly. However, in addition, Eastbourne had particular advantages. By the end of the century, a guide book described the town in the following terms:

Brighton is democratic; Hastings is salubrious (and, it must be confessed, a trifle dull); but Eastbourne, with its wide tree-lined streets, its miles of cultivated sea front with shaven lawns and prim *par terres*, is distinctly and decidedly elegant.

The 'elegance' of the town was ensured by the plan to which it had been built with the intention of attracting the middle classes to its streets and terraces of large houses.[3] Benjamin Soddy took advantage of all these amenities when he brought his family from South London to Bolton Road, one of the new terraces conveniently near the station in this respectable new town.

The Bolton Road house was not only well situated for the station, it was also just round the corner from Pevensey Road Wesleyan Chapel, attended by the whole family every Sunday.[4] Frederick Soddy always contended that this had a permanent effect on his character, resulting in the destruction of any religious belief and developing 'a plague-on-both-your-houses attitude toward religious controversy, later extended to social institutions in general'.[5] An uncomfortable side-effect was to inculcate 'the doctrine that truth is the main thing' and this in turn 'tended to make him too serious and to look on life's problems with too painful an eye'.[6] By 1919 he was suggesting that an adaptation of Keats' lines, 'Beauty is Truth, Truth Beauty, that is all/ Ye know on earth, and all ye need to know', to *Beauty and Truth and Duty*—that is all/ Ye need to know', would provide an inscription for the ideal university he was proposing.[7] This reaction to his family's religious beliefs was peculiar to Frederick. His brothers coped more easily, the next oldest, Thomas, becoming a Methodist minister.

When they moved, in addition to Benjamin and Hannah and the little boys, the household consisted of Lydia, the daughter of Benjamin's first marriage, and three live-in servants. These were Margaret Roberts, an elderly cook, Elina Roberts, probably her daughter, a fourteen-year-old general servant, and Mercy Relf, a twenty-four-year-old nurse. All seem to have come from Walworth with the family.[8] The two sons of the earlier marriage remained in London. Hannah's death just three years later had a distressing effect on them all. The children were still all under five and the servants, although probably willing, were unsuited to care for them. As a result, Lydia undertook the 'unsought and probably uncongenial burden' of mothering four small half-brothers, resulting in a change from being the daughter of the house to being 'the household dragon'.[9] For Frederick, at eighteen months of age the baby of the family, the effects were, at the time, catastrophic. He believed that they resulted in a childhood starved of affection leading to a speech impediment that was only cured during his school-days. This combination, he claimed, affected him very unfavourably and made him 'very self-sufficient and indifferent to his social surroundings'.[10]

Here he was exaggerating. While it was certainly hard to lose his mother at such an early age, and while perhaps his sister and father paid him less attention than he would have liked, there were compensations. Despite difference of rank and place the servants were devoted to him. Like many other children of his class, Frederick turned to 'below stairs' for support and affection. Margaret Roberts, the cook who had been with the family since before his birth, provided both. She, with the other servants, watched him perform scientific experiments, applauding the results and then clearing up after him in his nursery in the basement next to the kitchen.[11] In this she was to be the first in a whole series of women to whom Soddy turned for support during his life.

In addition the young Frederick was helped by his first school teacher at the local dame school, who earned his lifelong gratitude by curing him of the speech impediment that had affected his early childhood. At his second school, Frederick 'first tasted educated cultural influences',[12] and discovered an enjoyment of literature that would be a feature of the rest of his life and that later provided him with what F. A. Paneth described as one of his 'too numerous gifts'. As well as his skills as a scientist, 'he was such a good writer of English that it was all too easy for him to give his polemical essays the sting he wished'.[13] At this stage in his life, some kind of future in literature or writing might have been predicted for him but he moved on to his third 'more efficient but less refined' school at 6, Sussex Gardens, Eastbourne, under the tutelage of a Mr Hawkins, where he was to discover 'science'.[14]

He also began reading the essays of T. H. Huxley and, like H. G. Wells a couple of decades later, was so impressed that he might well have become a biologist.[15] Instead, he was taken to hear Sylvanus Taylor, Professor of Physics at Finsbury Technical College, lecture on electrical light and reacted so enthusiastically that he immediately set his ambitions on a career in chemistry or electrical chemistry. Consequently, plans were made for a different school, more experienced in teaching advanced science and mathematics, which would enable him to prepare for the City and Guilds College in Finsbury where he could qualify as an electrical engineer. The only suitable local establishment was Eastbourne College and so the sixteen-year-old schoolboy was enrolled there in the summer term of 1893.

Initially this was a problem. Frederick Soddy was far from the stereotypical public school boy of his time who loves sport, makes friends easily, and has a 'healthy disregard' for the benefits of academic learning. Although this philistine outlook had been modified by the nineteenth-century educational reforms within the public school system, it was never eradicated. Even if, by the late nineteenth century, the public schools, and the grammar and endowed schools that tried to emulate them, were beginning to aim at producing a 'gentleman, educated, courteous, well-spoken and considerate of others',[16] the older ethos was too strong. When Soddy attended Eastbourne College, '... the magazine [The Eastbournian] was usually filled with accounts of football, cricket and athletics matches . . .'[17] Soddy was of respectable, affluent, but firmly middle-class background which militated against the easy acceptance of these 'manly' but firmly 'aristocratic' ideas. He was also in some respects introverted. He disliked team sports of all kinds, preferring the solitariness of swimming in the sea off Eastbourne, 'starting with a boat at the Wish Tower, and swimming lazily along with the current, re-embarking perhaps at the Redoubt, some two miles away'.[18] Combined with his late arrival at the school, this clearly put him at some disadvantage, and might account for his lack of recognition there. He also remained there for only four terms, so had little time in which to make an impact.[19] It is perhaps significant that Eastbourne College, while celebrating its sportsmen and army connections, barely noticed it had produced a Nobel Laureate.

However, two things more than compensated for this. One of his near-contemporaries at school was Harold Cort Carpenter.[20] Carpenter was a couple of years older, shared Frederick's interest in science, especially chemistry, and was studying for a science scholarship at Oxford. He gained the Merton Science Postmastership in 1893. This success was made possible by the appointment of R. E. Hughes as science master at Eastbourne College. Hughes was Welsh and a young, enthusiastic Oxford graduate who had previously attended the University College of Wales at Aberystwyth. The appointment of a science teacher of such high calibre was somewhat unusual at the time. In 1872, the Devonshire Commission on Scientific Instruction and the Advancement of Science had reported:[21]

Nothing, however, can have much effect on the grammar schools and middle class schools of the country, generally, until the universities which give the key to education in the country, allocate a fair proportion of their endowments to the reward of scientific studies. Till such knowledge 'pays' at the universities, the middle class schools which look more or less to them, cannot be expected to change their course of instruction.

Frederick Soddy was therefore very lucky to find, barely ten years later, a teacher like Hughes whose 'major scholastic interest was in training boys for science scholarships at Oxford.'[22]

With Hughes's support and encouragement, Soddy's enthusiasm for his subject grew. The publication in *The Chemical News* in 1894 (when Soddy was only sixteen) of a paper in the joint names of R. E. Hughes and F. Soddy, on 'The action of dried ammonia on dried carbon dioxide'[23] was a material sign of talent. This led to the suggestion that instead of attending a City and Guilds college, Frederick might attempt an Oxford scholarship, following his friend Carpenter's example. However, Benjamin Soddy was less than enthusiastic. His aim was to see his children properly set up

The conclusion obtained from this is that the whole basis of economic theory needed to be rethought. If wealth was a flow which could not be saved, then the role of banks was called into question. This was of course a part of Soddy's programme for the future. An important part of his intention was to reveal the fraudulence of the banking system, but this was to come in the future. Immediately, only the barest outline was suggested:

> [During the war] the banks were bankrupt. The whole load of debt, accumulated in the prior piping times of peace and prosperity, had to be shouldered by the nation. With the productivity of the nation about to be fully absorbed in waging war it became bad debt. The national currency and credit, the absolute property of the nation, was pledged, free, gratis and for nothing, to the privately owned, dividend-producing banks to restore their solvency and save their clients from ruin.

While this article contains a very lucid statement of Soddy's beliefs at the time, it has remarkably little to do with National Guilds. At the time, the guild movement had become 'a force to be reckoned with'.[19] First formulated by S. G. Hobson in articles in *New Age* in 1908 and in his book *National Guilds* published in 1910, the Guilds had attracted the attention of 'the left wing intellectuals, especially . . . the young in universities'.[20] These 'rebels' had combined the ideas in Hobson's articles with the doctrines of Syndicalism and Industrial Unionism and, 'after failing to capture the Fabian Society . . . formed their own organisation, the National Guilds League'.[21] The Guilds had, by 1920, an impressive list of supporters. First and most important was A. R. Orage who, as editor of the influential journal *New Age*, gave the space necessary for publicity from 1908 until the 1920s. Of the others, G. D. H. Cole was probably the most influential, but members also included Clifford Allen, pacifist and Independent Labour Party politician; Arthur Greenwood, economist and academic; Frank Hodges, the miners' leader; George Lansbury, later leader of the Labour party; the philosopher Bertrand Russell, and R. H. Tawney, the historian.

The direct contacts between Soddy and this group are unclear. However, Cole had supported him as Labour Rectorial Candidate in 1920, as had Tawney, and the language of Cole's support suggested he had known Soddy for some time.[22] More importantly there is a broad similarity in overall aims between Soddy's writings and those of some 'Guildsmen'. This is particularly true in relation to the extent to which the 'basis of the Guild Socialist position was an attack on wages',[23] which was launched from a Marxist position. However, central to Soddy were the Ruskinian moral grounds also shared with the Guildsman. This can be summed up from *Unto this Last* where Ruskin brings to attention the outcome of production in his discussion of economics:[24]

> . . . it matters, so far as the labourer's immediate profit is concerned, not an iron

In the summer of 1920, and probably via the Labour Party, Soddy had come into contact with the National Guilds League and published an article in *The Guildsman: A Journal of Social and Industrial Freedom*,[18] which demonstrates the Ruskinian direction of his thoughts at the time. 'Economic "Science" from the Standpoint of Science' makes three main points, which are worth quoting at some length. The first of these is a discussion of energy as the basis of human life:

We now know how men live and move and have their being, so far as the economic aspect of that being is concerned. The moral and spiritual concerns of humanity are not comprised within that knowledge, but they are primarily based upon it, and if the basis is false the superstructure must ultimately collapse. Men derive their economic being entirely from the external inanimate universe, not from their own selves nor from any deity, malignant or well-meaning . . .

The economic being of men depends solely upon a continuous flow of energy from the inanimate universe applied by human intelligence and industry to human ends . . . this flow . . . is the wealth of the world . . . debt is a purely human invention and convention, and so also is usury or the spontaneous increment of debt at interest [emphasis added].

The second is the relationship of those who produce wealth, or harness the energy, to those who have been *assumed* to do so:

Creative institutions, such as scientific research laboratories, university and educational institutions, increase the real wealth of the world. Stock-exchanges, banks, financial houses, get-rich-quick concerns carry on the compensating process of increasing the world's debt.

. . . Usury can be justified only if we are committed to the position that the idle consumer of wealth is a superior creature to the producer of it . . . The multipliers of the world's wealth, the scientific investigator on the one hand and the manual workers on the other, have to beg from the philanthropic financier the laboratories and scientific apparatus and maintenance . . . There is not a University in the country which is not today indulging in the most appalling begging antics in public or private to raise money enough to continue to carry on its indispensable social work. There is not an industry in the country, however vital, the workers in which have not had to face the prospects of ruin and starvation to fight for a subsistence beyond the minimum of their animal requirements.

These propositions clearly demonstrate the broadly socialist nature of Soddy's thought at this period. However, his arguments are most noticeable for the synthesis of ideas about wealth and energy, and for uniting the cause of scientists and workers. His final point sounds the simplest but in fact was the most fundamentally challenging and connects most closely with Ruskin:

Wealth flows . . . the flow of wealth can only to a very limited extent be damned and then it does not bear interest, but on the contrary depreciates.

Soddy was then turning from his original researches on radioactivity under Rutherford, to concern himself with the economic and social problems this new source of energy might raise. The fact that Soddy was deserting the field for which he was professionally prepared to devote himself thereafter to political and social problems was perhaps the first example—and for a long time the only one—of a long-delayed awakening by the scientists to the possible social consequences of liberating atomic energy.

What is unclear is what Soddy had hoped to find in the Sociological Society. At the most general level, of course, support for his views on the relationship of science to society would have come from this group, but there is another, less obvious, link. This lies in the field of economics, the common factor running through Soddy's activities throughout the inter-war period. Beyond a too general appeal to the 'spirit of the times' it is difficult to know where this began. He had always prided himself on his materialistic outlook, he had long been aware of the importance of financial considerations in education and research and his Glasgow experiences had led to sympathy for Marxist analysis. What is missing is any sign of a particular moment or event which led to this new specific direction for his interests. However, a key influence was Ruskin. As Martinez-Alier writes, 'Ruskin was Soddy's favourite economist as he defiantly told an audience of economists in 1921.'[14]

In *Flors Clavigera*, and more particularly in *Unto This Last*, John Ruskin had produced an economic analysis of society in which he emphasised two particular features which would appeal to Soddy. Firstly, Ruskin emphasised the 'scientific' nature of economics, the necessity of an appeal to substantive, empirical evidence with concrete examples, which would have appealed to Soddy's professional practices. The first pages of *Unto This Last* are devoted to a dissertation of this kind into the difficulties of defining 'political science', with appeals to 'a perfectly logical and successful method of analysis . . . [for example] supposing a body in motion to be influenced by constant and inconstant forces . . . the disturbing elements in social problems . . . operate chemically.'[15] This is the kind of analysis which gave Ruskin's ideas so much influence in the Sociological Society and with Geddes, who wrote an appreciation of Ruskin in his 'John Ruskin: Economist' in 1884.[16] However, there was another theme in Ruskin's work which appealed more directly to Soddy. This was his application of the laws of physics to economics, and thus a definition of 'wealth' as 'a flow'; 'the flowing of streams to the sea [is] a partial image of the action of wealth . . . this wealth "goes where it is required". No human laws can withstand its flow . . .'[17] This perception was to underlie most of Soddy's own theorising. He was, as will be discussed in detail below, to apply the laws of thermodynamics to the flow of money and from this to develop his own, specific, theories of economics.

influencing the administration of Aberdeen University, but ended: 'the way the young men and women have responded and are responding to my lead here is most gratifying and amazing . . .'[10]

This letter is the only direct link between Soddy and Geddes. Despite marked similarities in their experiences and ideas there is no evidence of a closer acquaintance or further correspondence. However, it was probably partly through this connection that Soddy became involved in a consideration of the relationship between science and sociology and with the Sociological Society.

In November 1918, Soddy gave a lecture in Aberdeen on 'Scientific and Sociological'.[11] In this he discussed the

. . . meeting of two societies, the one primarily not interested in or concerned with social changes but powerless not to effect them.
The other primarily interested in social change and reform with the will to better the conditions of life, but without the knowledge and power to do so.

He went on to show the 'mutual distrust of the scientific man with society and the sociologist with science'. The rest of the lecture examines why this should be so, and suggests that the blame is unfair on both sides. Science had no responsibility for the use of its discoveries, it is a mechanistic subject, investigating the mechanisms of the universe for the sake of knowledge. Any discoveries made which impinge on the lives of the people should be the responsibility of the sociologist who should use them to move towards a more equitable and fair society. However, 'the war has brought to a climax a situation which was bound to arise', the questioning of the role of science, and so he asks, 'What is wanted' and answers 'A deliberate national consciousness of the opportunities which science—pursued in the first instance in the single-eyed interest of the Truth for its own sake—has put in the hands of men to mould their destinies'. 'The real use of science to humanity is so to increase the means of livelihood as to diminish the physical struggle for existence of society, not to increase it, to lengthen the hours for leisure, recreation and the cultivation of the graces and refinements of life, not to concentrate in the hands of a few all the leisure and put upon the backs of the many all the hard work.' Following this aim, Soddy began an investigation into why, as he saw it, scientific discoveries had been misused. This took him, in the first instance, to examine the role of the sociologist and to the Sociology Society.

In 1920 Lewis Mumford was spending some time in London where he was acting as editor of the *Sociological Review* and attending meetings of the Sociological Society at Le Play House, where he 'had occasion to talk with Frederick Soddy, whose little book on "Matter and Energy" in 1912 was one of my earliest acquisitions. (I still have it).'[12] Mumford's account continues:[13]

filing whether I employ him in growing a peach, or forging a bombshell; but my probable mode of consumption of those articles matters seriously. Admit that it is to be in both cases 'unselfish', and the difference, to him, is final, whether when his child is ill, I walk into his cottage and give it the peach, or drop the shell down his chimney, and blow his roof off.

Soddy's association with both the National Guilds and the Sociological Society was brief. He disappeared without trace from the records of the Sociological Society, reappearing some twenty years later in one of its junior organisations, the Le Play Society. His connection with the National Guilds was more abruptly severed. Having published the article quoted above, he was advertised as one of a series of speakers on 'What I Think of National Guilds' at a meeting chaired by G. D. H. Cole in October 1920. Soddy did not appear at the meeting, which was replaced with a lecture by Cole, and although no explanation was given we can only assume that given his other 'interests' this was Soddy's 'public' break with the Guilds.[25] However, Cole's and Soddy's paths were to cross again and New Britain included many, especially J. T. Murphy and Ben Tillett, who had been involved in the National Guilds.

Soddy appears to have spent the latter part of 1920 and most of 1921 in a period of reading and thinking. This resulted in a thoroughgoing reassessment of his allegiances. As we have already seen, he associated himself most publicly with the Labour Party up to 1921 when he stood as Labour Rectorial Candidate at Aberdeen and he was closely involved with the National Union of Scientific Workers in 1920. After 1921 he gradually severed these connections, although he continues his association with the Labour Party until at least the end of 1921. It is also interesting that, as late as 1924, he was still lecturing to socialist groups.[26]

More positively, this period marks the beginning of his work as a monetary reformer. During February 1920 he was making notes on Ruskin's essay 'Veins of Wealth' from *Unto This Last*, together with Volume 1 of Capital and H. D. Macleod's *Theory of Credit*.[27] He developed this pattern of study, moving on to the monetary reformers Silvio Gesell and Arthur Kitson, with whom he agreed, at least to a limited extent, and J. M. Keynes, Major Douglas and A. R. Orage, with whom he disagreed. It is probable that he was influenced, initially at least, by reading *New Age* which, under the editorship of Orage, had in the pre-war period 'reached the position of being compulsory reading for anyone who claimed to be of the Left in literature, art, drama, politics, economics and what-not.'[28] A sense of its importance can be gained from a list of contributors from just before the war. Orage wrote 'Notes of the Week'; J. M. Kennedy, at one time a journalist on the *Daily Telegraph*, contributed 'Foreign Affairs'; G. D. H. Cole, Rowland Kenney (the first editor of *The Daily Herald*), and S. G. Hobson wrote on politics. J. A. Hobson and Arthur Kitson wrote on economics and in addition there were a number of foreign correspondents. The list of writers concerned with the arts is even more impressive. T. E. Hulme, E. Belfort Bax, A. M. Ludovici and J. M. Kennedy wrote on recent trends in philosophy; Conrad Noel, Allen Upward and Orage contributed to discussions on religion; Walter Sickert was an arts reviewer. Others like Beatrice Hastings and A. E. Randall combined a number of roles.[29]

While retaining a broad spread of topics in the journal, Orage had become increasingly interested in the 'New Economics' and had collected his own writings on the subject which he published in 1917 in *An Alphabet of Economics*, in which he paid tribute to Ruskin.[30] He had also begun publishing articles on monetary reform, firstly, from 1908, by J. A. Hobson and then, from 1912, by Arthur Kitson and finally, immediately after the first world war, by Major C. H. Douglas. Kitson had written *A Scientific Solution to the Money Question* in 1894 in which he argued for an increased money supply and governmental control of credit, while Douglas was the founder of Social Credit, which was the most widely supported proposal for monetary reform promulgated in the 1920s. From 1918, Social Credit had been explained and publicised in the pages of *New Age*, generally replacing Guild Socialism which in any case had 'died' by 1921.[31] Social Credit, was, according to an early supporter, the Scottish poet Hugh MacDiarmid,[32]

... 'economic democracy' because by socializing credit it aims to decentralize financial initiative (to decentralize it down to the individual in fact and establish the economic independence of the individual). In the light of this policy the aim of industry is simply to supply goods to people who want them.

At a more theoretical level Douglas insisted that 'the purchasing power' of individuals must consist of *both* wages, salaries and dividends—what Douglas called 'A payments', *and* elements of other payments to cover raw materials, bank charges and other external costs—what he called the 'B payment'. Both these elements had to exist in all economic calculations, hence the theory of social credit rested—actually rose or fell according to Douglas—on the A+B Theorem. Additionally, Douglas argued, the twentieth century was the age of post-scarcity—the age of leisure. It was only the puritanism of an earlier and less advanced age which insisted that workers be rewarded only according to their 'effort'. In the Douglas system every individual, regardless of age, skill or effort was to be given the National Divided, the ultimate and 'logical successor to the wage'.[33]

The first fruits of Soddy's reading in the area of monetary reform emerged in two lectures he gave to 'The Student Unions of Birkbeck College and the London School of Economics, November 10th and 17th, 1921', chaired by Sir Richard Gregory, editor of *Nature* and still, like Soddy at this point, supporting the Labour movement in a battle for resources for scientific research.[34] The lectures were then published in the pamphlet *Cartesian Economics* in May 1922.[35]

There are a number of important points about these lectures.[36] Firstly, there was the fashionable appeal to science. Mirowski has shown that 'the importation of physical metaphors into the economic sphere has been relentless, remorseless and unremitting in the history of economic thought' throughout the nineteenth and twentieth centuries.[37] In line with this tradition, Soddy used science for his model, but he went further than just a metaphor. It was his 'intention to try to bring the existing knowledge of the physical sciences to bear upon the question "How do men live?".'[38] He attempted to answer this question by applying the first and second laws of thermodynamics to what he regarded as the fundamental point, which is that life depends 'on sunshine'. In addition, he was seeking to explain what he regarded as the central problem of economics,

the interaction of ... physics and mind in their commonest everyday aspects, matter and energy on the one hand, obeying the laws of mathematical probability or chance as exhibited in the inanimate universe, and, on the other, with the guidance, direction and willing of these blind forces and processes to predetermined ends.

Here he was combining two kinds of economic theory, the scientific and the moral—the tradition of Jevons with that of Ruskin. The importance of his approach was that it insisted on what may be described as the 'political', on the relationships between the individuals in society and their economic lives. The result was not always successful, as in the infelicitous phrasing:[39]

The principles and ethics of human law and convention must not run counter to those of thermodynamics. For men, no different from any other form of heat engine, the physical problems of life are energy problems.

However, the essential point, of the need to consider the duality of human life, led to an intricacy in his theorising which has enabled different readers to seize on different points as essential.

Soddy argued that all life depended on energy. All energy came ultimately from the sun and could only be converted to a form which could be used by animals by the actions of plants. The tendency of this energy is entropic, it moves only in one direction, to imperceptibly warm the oceans. Periods of prosperity in the past have depended on finding new sources of power and on new discoveries. The use of coal depended not only on the sinking of mines, but on the discovery that using it for fuelling machinery could develop new kinds of power and technical achievement. Whatever discoveries might be made, there would still be a dependence on sunshine, on a revenue of energy. Energy in a form which could be utilised for life could not be stored. Bodies need food and this food ultimately decays, therefore it must constantly be replaced and no discovery had (has?) yet been made which can replace the plant in this process, which Soddy labelled 'life-use'. Populations come to the limit of expansion of food supply, even after attempting to expand this through colonisation and factory manufacture and artificial fertilisers. This leads to a race for markets for the produce of factories which results in reducing capacity for food production. At this point, he concludes 'the aspect of this mad system which is now uppermost in the minds of many thoughtful people, its inevitable end in world-war.' Soddy finally (mis)quoted Ruskin's classic statement:

Capital is a root, which does not enter into vital function till it produces fruit. Capital which produces nothing but capital is root producing root, bulb issuing in bulb, never in tulip. The Political Economy of Europe has hitherto devoted itself to the multiplication of bulbs. It never saw nor conceived such a thing as a tulip. Nay! boiled bulbs they might have been, glass bulbs—Prince Rupert's drops consummated in powder—well if it were glass powder and not gun-powder.[40]

To emphasise his point, he added, 'I should like to see a memorial to Ruskin bearing these words erected on the plains of Flanders'.[41]

Soddy's suggestions for the future are less clear, at this point, than his analysis of the faults of society. A part of his ideas was to increase production and discovery, and by this means the wealth of the society. This is another of his insights from Ruskin. He quotes Ruskin to show that 'what one person has, another cannot have'. Increased production would ease this situation, and especially if equity could be introduced into the distribution, since 'of what use are the discoveries of scientific men into new modes and more ample ways of living so long as the laws of human nature turn all the difficultly won wealth into increased power of the few over the lives and labours of the many?' This led to his most original point which was to distinguish wealth and debt. He was to elaborate on this over the years but here he merely asserts the difference before embarking on a short discussion of the nature of money.

Over the next three years Soddy developed these ideas, reading more widely and lecturing to various audiences. For example, he lectured 'in Glasgow and Edinburgh . . . under the auspices of the Union of Democratic Control, Edinburgh Peace Council, and Glasgow Socialist Teachers' Society.' In these towns he lectured on 'Physical Fallacies That Breed War', 'The Inversion of Science and its Consequences' and 'Science Under Humane Government as the Great Emancipator and Peacemaker'. 'The lectures were delivered with Professor Soddy's customary forcefulness, cogency and simplicity, and we hope that they will be published in pamphlet form . . .'[42]

These hopes were fulfilled as Soddy produced a more detailed version of his proposed scheme for the use of scientific discoveries for the improvement of society in his pamphlet *The Inversion of Science and a Scheme of Scientific Reformation* in the same month, January 1924.[43] As he announced in the foreword, Soddy considered that his main principles of reform were derived from the work of Gesell and Kitson. Although he had read the work of Douglas, and had 'been stimulated by his work on credit reform and on the inefficiency, sabotage and waste that at present are inseparable from our system of distributing wealth', he preferred his own system of reform, 'the simpler scheme of basing currency on index number'. This was 'at once capable of being understood in terms of physical reality and powerful enough to liberate democracy forever from the usurpation of the usurer.'[44] The development of these 'simpler ideas, appreciated by the Social Credit movement over the last two years',[45] were to contribute to a split in the Social Credit organisation at the second annual conference in 1925.

In his 1924 pamphlet, *The Inversion of Science*, Soddy based his proposals on two developments from *Cartesian Economics*. Firstly, he returned to the laws of physics to show that just as perpetual motion is impossible in the physical world, so should the notion of perpetual interest on investment, and most especially perpetual compound interest, be in the economic world.[46] However, more importantly, he examined what he considered to be the essential cause of the economic ills of the country which led to poverty and in turn to war. This was the 'money system' which had a number of faults, most of which stemmed from the banking system which allowed fractional reserves, loans based on securities which were never sold, and again, a misunderstanding of the true nature of wealth. In addition, there was a shortage of money circulating in the system which resulted in a lack of purchasing power which, in turn, led to a reduction in production and to the destruction of food in a country where the poor were starving.

The solution to these problems was complicated but consisted of two essential features. Firstly, the banks should be nationalised so that they did not need, indeed were prohibited from, making a profit. Secondly, and this was the central plank of Soddy's platform, gold as the measure of value of money should be replaced by an index number. This number could be related to almost anything but 'as the chief object is to ensure a sufficiency of necessities before luxuries, the budget of a working class household would be a sufficiently satisfactory basis. The relative amounts of all the commodities consumed are taken into account in fixing what is known as the index number.'[47] A board of statisticians would be responsible for fixing this number and then the amount of money in circulation could be adjusted accordingly. More money would be issued as prices fell and withdrawn as they rose again. This leads to the connected point which is Soddy's belief in the need for a plentiful money supply for national prosperity. He continued the underconsumptionist analysis of Ruskin, of Douglas, and of his own earlier work to argue that the existing difficulties were due to a lack of goods bought in the market due to a desire to save money. In turn, underconsumption led to underproduction which in turn led to unemployment. This formulation was unlike that of Keynes, who suggested that state-funded capital investment was the solution to underemployment. Soddy, like Gesell, insisted that only an increased amount of money in circulation, either in total quantity or in speed of circulation, would rectify the situation. With a greater money supply, more goods would be bought, more would need to be produced. More work would lead to less poverty.

There is a final theme in this work which was to lead to what has been seen as both the most original and the most opaque part of Soddy's monetary theory, the notion of a distinction between 'real' and 'virtual' wealth. Here, he talks of real and fictitious loans. To simplify, a real loan was one for which something was sacrificed—shares sold by the bank, for example; a fictitious loan was one for which security was just the share certificate. In the latter case the shares remained in the possession of the bank which, under the then obtaining system, lent non-existent money through a cheque or similar mechanism, without actually suffering any deprivation or reduction in its reserves. To Soddy, 'virtual wealth' became the trap, the central problem of monetary theory.

Soddy finally listed 'some possible objections'.[48] The first of these was the effect of saving, which would reduce the quantity of money in circulation. The solution to this was simple: more money would need to be printed to maintain the quantity of money, 'the individual may save but the community cannot'. This need not be a problem since 'if science continues to develop there is no limit in sight to the community's expansion' which would in turn meet future calls on the revenue. Simply employing the unemployed without increasing consumption would be a step in this direction.

The second objection is interesting since it reflects back on the more humanist concerns of the earlier pamphlet. It is 'the bogey of overpopulation'. This is based on the facile argument that 'the human population, like that of rats, always tends to increase to the limit of the food supply'. The response to this is that 'fecundity is favoured by poverty and diminished by prosperity'. He goes on to turn the eugenicist argument on its head by suggesting that, if the problem was the tendency of the poor to reproduce faster than the wealthy, or the working faster than the middle class, then the solution lay in education and improved environment. 'As regards to the highest type of all, genius is as apt to appear in one cradle as another, and even the eugenicist would have difficulty in tracing pre-eminence to heredity.'[49] The conclusion to this train of thought is that 'if the object of society is to breed cannon-fodder for fratricidal wars, it cannot keep the people too poor . . .' This, like the fact that many of the groups Soddy addressed in the early 1920s were pacifist or quasi-pacifist, points to his continued and continuing fear of war. Like many who had been traumatised by the events of 1914–18 it was not so much that specific events led Soddy to a fear that conflict was about to break out, rather that a generalised horror created by the memory of the conflict constantly informed his life and work.

It is necessary now to return to the more overtly political aspects of the monetary movement. By the time *The Inversion of Science* was published in January 1924, support for Social Credit had grown but signs of splits in the movement were appearing and differences of opinion about policy were emerging. The movement had already overcome some setbacks. In 1922 it had lost two kinds of support. Most importantly, the Labour Party Committee which had been set up to examine Social Credit rejected the proposals. Secondly, Orage went to the Gurdjieff–Ouspensky school at Fountainbleau, leaving the journal in the hands of the unsympathetic Major William A. Moore, prompting Kitson to protest, 'In common with many others . . . I have been somewhat startled to see the change in the tone of *New Age* since its change of ownership.'[50] However, by June 1923, Moore had retired and had been replaced by Arthur Brenton who had been running *Public Welfare: A Review of Contemporary Finance and Industry* which was dedicated to 'Promote the Expansion of Citizen Purchasing Power by the Credit and Costing Methods Discovered and Announced by Major C. H. Douglas and the *New Age*'. *Public Welfare* in turn became *Credit Power*, 'A Critical and Constructive Review of Financial Policy from the Standpoint of the "New Economics."'[51] Finally, *Credit Power* was incorporated into *New Age* when Brenton took it over. Throughout this period grass roots support of Social Credit, or its simpler versions, was encouraged through local study groups.[52] The heterogeneous nature of their beliefs is, however, demonstrated by the appeal made in 1923 for joint action between all monetary reform groups regardless of their precise theories.

From 1924, a series of proposals was made suggesting centralisation of some kind. This came to a head in 1925, at the second national conference at Swanwick. At this conference the Sheffield Group argued that there was enough general support to press for the formation of a 'more organised and purposeful movement'.[53] This move was opposed by London members, perhaps because they had already experienced one failure through the National Guilds. The Sheffield group prepared an agenda for the conference and, having been promised the support of eighty-seven Social Creditors, six qualified supporters and only three opponents, assumed that their views would prevail. 'However, in some way which is not clear, the conference repudiated the prepared agenda'[54] and thus destroyed the chance of Social Credit being an organised 'party'. The attack on the Sheffield motion was masterminded by W. T. Symons, one of the London members and Chairman of the Hampstead branch of the ILP, who had come into contact with Social Credit through Orage and *New Age* in 1916. Symons will be of more importance in the following chapter where his connections with Soddy will be explained. At this point, though, they were on opposite sides.

There is no way of knowing for certain at this stage what, if any, connection Soddy had or had had with the loose grouping around Douglas's Social Credit or how he came to be at Sheffield, although his work had been reviewed in the social credit journal *Public Welfare*[55] and then eighteen months later the same journal, by now called *Credit Power*, reported his lectures.[56] However, both he and Kitson lectured to the conference and they and fifteen others then held a special business meeting, separate from that organised by Symons, to discuss the Sheffield Group's proposals. As a result of this meeting, Soddy, with Kitson, was amongst the twenty seven supporters of the Sheffield proposal who, when the agenda was repudiated, withdrew from the conference to form their own, separate movement, the Economic Freedom League.[57] Others involved included a Dr Rice who headed the new movement and H. E. B. Ludlam who was to edit the new movement's journal, the *Age of Plenty*.

The result of the split was two separate, but not antagonistic, kinds of social credit. However, the grounds of the breach were not only organisational. There was some suggestion that Douglas himself was deliberately obstructing the development of a more structured organisation, but the central difference of opinion centred on an appeal to Soddy's alternative, simpler monetary theory which did not rely on Douglas's A+B Theorem.

The subsequent development of these groups demonstrates the appeal of the different kinds of monetary reform on offer. One apparently unlikely source of support was John Hargraves's Kibbo Kift Kin. John Hargrave had set up the Kin in 1920, with Patrick Geddes and H. G. Wells amongst its advisory councillors and it reached its maximum membership of just over two hundred in 1924.[58] In that year Hargrave retreated to the Welsh mountains to consider the lack of success of the Kibbo Kift Kin with Rolf Gardiner, editor of *Youth* and founder of Cambridge Social Credit study circle. The result of this consultation was an instruction in *Nomad*, the Kin journal, that members should read both Douglas and Soddy.[59] After the Swanwick split, Hargrave supported both Soddy and Douglas in turn until finally, in 1928, he was the main speaker at the Economic Freedom League conference. Even after this the Kindred were urged to take *New Age*. Finally, Philip Kenway, Hargrave's front man, suggested in letters to *New Age* that a third way between Social Credit and the Economic Freedom League could be found. This was turned into a call for mass action, the 'Third Line', by Hargrave and an intermediate group, and the Economic Party was set up.[60] This party lasted only until 1930 and after this it metamorphosed into first the Crusader Legion, or the Iron Guard, which then reunited with the remnants of the Kibbo Kift Kin to become the Green Shirts, whose main attraction was their uniform and marches and an appeal to the individual. Nevertheless, for the first time clear connections were made between Social Credit and fascism.[61]

This link, which had some basis in theory, continued to dog Social Credit and other kinds of monetary reform. Most obviously the connections were personal and intellectual rather than organisational. Rolf Gardiner was later a member of the British Union of Fascists and was interned during the Second World War. Lord Tavistock, heir to the Duke of Bedford, for example, was a leading figure in the National Credit Association, which held a conference in 1933, but he was also certainly on the far right and was thought to have connections with the British Union of Fascists.[62] Similarly, by 1933, Ludlam, in the pages of the *Age of Plenty*, had replaced Social Credit with fascism, soliciting subscriptions for the British Union of Fascists.[63] More importantly, Sir Oswald Mosley's foundation of the BUF was a part of his attack on the monetary policies of both Labour and Conservative governments after 1923.[64] He had been connected with both Douglas and the *Age of Plenty* in the 1920s and his policies as Chancellor of the Duchy of Lancaster in 1929–30 have been seen as 'an unsuccessful attempt to produce a programme which would merge a synthesised version of . . . the political machinery of the establishment' to the rejected knowledge of the economic underworld, the monetary reformers.[65] There were also connections between Ezra Pound and monetary theory. Soddy has, as a result of this, been accused of fascist sympathies.[66] This charge is one which Soddy and others constantly rejected throughout the thirties, arguing that to the non-left the organisation was seen as socialist, leading to wealth and security for all. Soddy believed his beliefs were neither fascist nor communist, but provided a third way in between. This is supported by a careful reading of *New Britain* weekly, whose tone and contributors were largely of the left, even if there are occasional signs of a generalised sympathy for aspects of the right. For example, in May 1934, Winifred Horrabin, certainly not a figure of the right, wrote of 'the welter of tragedy that is modern Germany under the Hitler Reich'.[67]

The organisations of Social Credit remained confused. In 1932, the West

Riding Douglas Social Credit Association was founded, and its leading member was C. M. Hattersley, once prominent in the Economic Freedom League. At the same time a new journal, *This Prosperity*, edited by Robert J. Scrutton in Coventry, published the suggestion of a national petition to force the issue of Social Credit on a national scale through the demand for royal commission. This was to lead to the formation of the Petition Council in 1936 which will be discussed in more detail below.[68] After 1935, when their electoral campaign failed, 'Social Credit purity began to be watered down . . . Soddy and Kitson became more prominent' until 1938 saw both the last issue of *New Age* and the end of Social Credit as an active movement.[69]

While he was involved in these movements, Soddy continued his writing and in 1926, his major work on monetary theory, *Wealth, Virtual Wealth and Debt*, was published.[70] Its frontispiece is a quotation from *Unto This Last* and it was dedicated to Soddy's other hero, Arthur Kitson, 'the British pioneer of the new economics to whose writings the author owes his initial interest in the fascinating problems of wealth and currency'. This was in essence a more detailed and complete statement of the ideas which Soddy had been developing since the war, but with the addition of two new ideas. He firstly returned to the question behind all his theories which was how the riches promised by science could be distributed in such a way that prosperity for all would result.[71] He then explained that the basic problem, the 'economic paradox' of his title, lay in the necessity for a 'genuine permanent surrender of rights to consume . . . the *genuine* initial abstinence'.[72] However, it was his second and unique formulation which attracted the most attention. This was his theorising of 'virtual wealth'. Virtual wealth was a part of Soddy's desire to stabilise the value of money. Virtual wealth was a theoretical concept, it was 'the purchasing power of the £1 sterling . . . divided by the total quantity of money'. The difficulty was then the purchasing power, but this could be settled by recourse to an index number. He argued that in a society where other weights and measures had some kind of 'absolute value', it was absurd that money should be free-floating. In more detail than before, he suggested an index number 'that so expresses the money cost of the quantities of the things required in due relative proportion to maintain an average family . . .'[73] which could be fixed and to which the value of money could be related.

Perhaps as a result of his experiences with the Social Credit movement, and the desires shown there for some kind of concrete proposals for action towards 'doing something', Soddy provided a list of twenty two 'practical conclusions'. Amongst these, reformation of the banks and banking system loomed large; replacing the gold standard with the index number and fixing the value of money were the other important elements. Unless some such scheme was developed, 'it is only a matter of time before another war will come, greater and more terrible than the last . . .'⁷⁴ His conclusion then reverts to the kind of elevated, almost mystical language which seemed always to be a part of Soddy's explication of areas in which he had a deep emotional attachment. Just as his message about the dangers of radioactivity before the war had been couched in confusing, sybillic, poetic phrases, his final paragraph here is somewhat extraordinary for an economics text. He wrote:[75]

The wheels of God grind small, but they grind exceeding slow. O future! we the dying salute thee! The course is set, the race is nearly run . . . Give us back, O powers of the light! but one more hour before the pendulum of night descends again. The lamp is lit, but its beam needs time to grow ere those to come can hope to grope their way. Slow down the sunset and upspeed the dawn, lest youth resurgent should arrive too late.

In case this was not persuasive, he then rephrased Milton's 'On the Recent Massacre' to refer not to religious but monetary persecution leading to war, slaughter and scattered bones!

The influence of *Virtual Wealth* spread far outside the narrow confines of English Social Credit. It was particulary important in the United States of America.[76]

In the United States, the group known as the Technocrats became interested in Soddy's plan during the depression following 1929. They bought so many copies of his 1926 edition of *Wealth, Virtual Wealth and Debt* that a new American edition was brought out in 1933.

The Technocrats had a relatively short life, dying away 'rather quickly in a welter of internal recrimination and nonsense'.[77] However, through their influence Soddy's work was taken up by more mainstream American economists. A group of economists at the University of Chicago picked up on his '100 per cent Reserve Scheme', while similar ideas were extremely important to Professor Irving Fisher of Yale.[78] Soddy also influenced the maverick Labour politician, Thomas Johnson. In the aftermath of the fall of the Labour Government in 1931,[79]

Johnson touted in particular the economist Professor Frederick Soddy whose work had inspired his War Loan proposal and who, in Johnson's view, was one of the few authorities on economics who grasped the importance of the National Debt in the country's financial economy.

Despite the gloom expressed at the end of *Virtual Wealth*, Soddy returned to his programme of reading, lectures and publications. In January 1927, he lectured to the students of the University of South Wales and Monmouthshire on 'The Wrecking of a Scientific Age'. This became the third of his series of pamphlets, *The Wrecking of a Scientific Age*.[80] In his introduction to this lecture Soddy demonstrated the pessimistic direction of his thoughts by asking, 'is there a man or woman in this room who really believes that "the war to end war" did end it, or, that in the fullness of time and the more certainly the longer delayed it is, another will not be upon us which will imperil the whole future of civilisation?'[81] He then repeated his by now familiar account of the contrast between scientific and technical discoveries since the steam engine of James Watt and the inauguration of the 'science' of political economy by Adam Smith and the decline of society, to the point where[82]

workpeople have become degraded and C3 physically, herded in foul slums, their livelihood rendered increasingly precarious by competition with mechanical labour-saving appliances of every description, periodically condemned, as now, to long periods of forced unemployment and destitution, able and willing—never more so—to make almost any conceivable sort of wealth humanity can desire and not allowed to make enough to keep the wolf from their own door.

He blamed this, again, on the situation where the working classes produced goods so that the *rentier* class could live on rents, profits and, above all, interest on debts. Emphasising the moral dimension of what he said, Soddy repeated his view that science could do away with all debt and replace poverty with prosperity.

Soddy's fourth and last pamphlet in this series was the text of a lecture which he gave to the students of the University College of Wales, Aberystwyth, in October 1927.[83] *The Impact of Science on an Old Civilisation* is very much more general than Soddy's earlier pamphlets. It contains no new ideas, but, like the conclusion of *Virtual Wealth*, returns to his anxieties of the pre-war decade—the possibility of another war. He emphasises the importance of a new economics because otherwise[84]

... the antagonism between science which makes wealth plentiful and a civilisation based on the principle of scarcity produces congestion and deadlock in the whole economic system ... [Without reform, the only solution] is simple, it is devastatingly effective, but it is dangerous to the continuance of the civilisation which yet cannot exist without it. It is War.

After the publication of *The Impact of Science*, there was a period of some years during which Soddy wrote relatively little. The precise reasons for this are unclear, but it seems likely that he, like many of the other monetary reformers, was forced to rethink his position. This was partly because of national economic developments. The middle of the decade was marked by two events which particulary affected the reformers. The General Strike and the return to the gold standard marked a watershed in the affairs of the 1920s. The effects of these events have been described in detail in many recent works,[85] but were summarised by A. J. P. Taylor:[86]

The half-hearted gold standard was a symbol of the postwar restoration. To outward appearance, all troubles past, all passions spent: Free trade, the gold standard, stability at home and abroad . . . Baldwin's second government passed five quiet years, with one alarming, and perhaps unnecessary, interruption: the General Strike.

Although the contraction of British Industry was to continue, and politics and the economy were only apparently tranquil, any sense of crisis had passed with the return to work of the miners, to return only at the very end of the decade and the beginning of the thirties. Ronald Blythe described the period between as the one occupied by 'dancing in Britain and America, and sunbathing in Germany', it was pre-eminently the age of the Charleston.[87]

For Soddy, the events of the early twenties challenged his beliefs and aims and sent him in a somewhat unexpected direction in the thirties. His most cherished dream, the exploitation of science towards a more prosperous and equal society, had been destroyed. Rutherford had shown how elements could be disintegrated by bombardment by alpha particles,[88] but there was no sign of any advancement in the study of the release of energy through that means. His particular part of atomic science, empirical radiochemistry, was temporarily without any obvious direction while the development of theories in atomic physics was all-important.[89] His father-in-law, Sir George Beilby, with whom he had so much in common, had died in 1924 and his company, the Cassel Cyanide Company, was about to be absorbed into ICI,[90] thus removing the possibility of the kind of close local collaboration which Soddy and Beilby had effected during Soddy's Glasgow days. His dreams of a communistic science, organised through the Union of Scientific Workers and National Guilds and funded by a socialist labour government, had come to nothing. The Union had become the non-political Association of Scientific Workers; the Labour Government, despite promises made in their plans for Reconstruction, had actually done nothing to advance academic science. The National Guilds Scheme had also faded, to be replaced largely by Social Credit which in itself left something to be desired.

Even this seemed to have got nowhere. Social Credit had been rejected by the Labour Government in 1922. Monetary theories had had so little effect that amongst supporters of the various schemes a belief grew in a conspiracy of silence which occasionally verged on paranoia. For example, the poet Hugh MacDiarmid described in the early thirties how[91]

. . . the powers responsible for keeping all mention of his name [Major Douglas] and discussion of his ideas out of every paper and periodical except those little-known periodicals of extremely limited circulation which existed for the specific purpose of promulgating Douglasism, for a period of some ten years have at last decided to 'lift the ban'.

The result of these disappointments is summed up by the change in Soddy's political beliefs. At the end of the war he had been towards the left of the Labour Party, supporting the ILP and looking forward to a socialist state. His pamphlets of the early twenties were published by Henderson's, of the Charing Cross Road, known as 'The Bomb Shop' because of the revolutionary nature of its stock and publications, and were also available from the ILP bookshop. The extent of his move away from Labour can be demonstrated in an exchange over *The Impact of Science*. This was reviewed by John Strachey in 1928 in *The Socialist Review*, under the title 'Soddy Goes Socialist'.[92] He began: 'Socialists should read Professor Soddy's latest pamphlet, *The Impact of Science on an Old Civilisation*.' Strachey continued, putting this in the context of his own development from a currency crank, who believed that the cause of all problems was lack of currency, to realise that the problem was deeper than this.[93] According to Strachey, lack of currency was only a symptom of the problem with modern capitalism, 'to think, as I used to think, that one could have a scientific socialised banking system, pursuing a rational, scientific currency policy, whilst leaving the rest of capitalism and class domination intact, is mere Utopianism.' Strachey suggested that what he saw as Soddy's new position could be summed up in one sentence which he quoted in italics for emphasis: '*If science were to make everyone well off, the present source of the unearned income of the rich would be completely dried up.*' Unfortunately, other parts of the pamphlet contradicted this, but Strachey felt they were only the leftovers of Soddy's earlier way of thinking and that in time Soddy would realise that the only sensible conclusion to his thinking would be for him to join the socialist party as an alternative to 'throwing his little orange pamphlets at the great barred doors of Capitalism'.

However, Soddy explained in his answer that this was actually the reverse of the truth.[94] Since the war, Soddy said, his views 'as to the true solution of the social problem . . . have *grown* in the past few years since the War, but . . . away from Socialism and towards "currency reform".' His conviction that the ills of society could be blamed on the currency system had grown, as had his distrust of banks. At the same time, however, his distrust of socialism as a cure had grown. While he 'fully accept[ed] the doctrine of economic determinism in history . . .' this did not extend to supporting 'the Marxian Party'. Soddy argued that the existence of poverty was just as essential to the success of the socialist party as it was to the capitalist. 'One can hardly attend a Labour meeting without encountering the most hopeless type of reactionary, compared with whom the true blue Tory is nowhere. Never tired of killing Capitalism with his mouth, he is equally the bitter opponent of every proposal for fundamental reform . . .' Just as much as the capitalist, the socialists needed to retain the *status quo*.

The period of calm was spectacularly shattered by the Wall Street crash of 1929, by which time Soddy had found another intellectual home.

Notes and references

1. Aldcroft, D. H. (1970). *The Inter-War Economy: Britain, 1919–1939*, p. 35. Batsford.
2. Winch, D. (1969). *Economics and Policy*, p. 67. Hodder and Stoughton.
3. Marwick, A. (1991). *The Deluge: British Society and the First World War*, p. 329. Macmillan.
4. Martinez-Alier, J. with Schlupmann, K. (1987). *Ecological Economics: Energy, Environment and Society*, p. 143. Basil Blackwell, Oxford.
5. Martinez-Alier, J. with Schlupmann, K. (1987). *Ecological Economics: Energy, Environment and Society*, pp. 127–44. Basil Blackwell, Oxford.
6. Meller, H. (1990). *Patrick Geddes: Social Evolutionist and City Planner*, p. 2. Routledge.
7. Meller, H. (1990). *Patrick Geddes: Social Evolutionist and City Planner*, p. 3. Routledge.
8. They are quoted in Boardman, P. (1978). *The Worlds of Patrick Geddes*, p. 305. Routledge and Keegan Paul.
9. Quoted in Stalley, M. (1972). *Patrick Geddes: Spokesman for Man and the Environment*, p. 82. Rutgers University Press, New Brunswick.
10. Strathclyde University Archive, T-GED 9/1447.
11. Professor Frederick Soddy Papers, Bodleian Library, Oxford, MS Eng Misc b179.
12. Mumford, L. (1979). *My Works and Days: A Personal Chronicle*, p. 72. Harcourt Brace Jovanovich, New York.
13. Mumford, L. (1979). *My Works and Days: A Personal Chronicle*, p. 72. Harcourt Brace Jovanovich, New York.
14. Martinez-Alier, J. with Schlupmann, K. (1987). *Ecological Economics: Energy, Environment and Society*, p. 132. Basil Blackwell, Oxford.
15. Ruskin, J. (1968). *Unto This Last*, p. 115. Everyman Edition, Dent.
16. For discussions of the effects of Ruskin on Geddes's economic theorising see Finlay, J. L. (1972). *Social Credit: The English Origins*, Chapter 1 *passim*. Queen's University Press, McGill; Meller, H. (1990). *Patrick Geddes: Social Evolutionist and City Planner*, pp. 29–30. Routledge.
17. Ruskin, J. (1968). *Unto This Last*, pp. 147–8. Everyman Edition, Dent.
18. The following quotations are all from Soddy, F. (1920). Economic 'Science' from the Standpoint of Science. *The Guildsman: A Journal of Social and Industrial Freedom*. July 1920, pp. 3–4.
19. Finlay, J. L. (1972). *Social Credit: The English Origins*, p. 76. Queen's University Press, McGill. Except where otherwise stated, the history of the National Guilds and their relationship with Social Credit described here relies on Finlay.
20. Cole, Dame Margaret (1971). *The Life of G.D.H. Cole*, p. 52. Macmillan.
21. Cole, Dame Margaret (1971). *The Life of G.D.H. Cole*, p. 52. Macmillan.
22. *The Lord Rector*. Aberdeen University Labour Club Magazine. 19 October 1921, p. 8.

23. Finlay, J. L. (1972). *Social Credit: The English Origins*, p. 77. Queen's University Press, McGill.
24. Finlay, J. L. (1972). *Social Credit: The English Origins*, p. 184. Queen's University Press, McGill.
25. *The Guildsman*, 27 October 1920.
26. *Credit Power*, January 1924.
27. Professor Frederick Soddy Papers, Bodleian Library, Oxford, MS Eng Misc b175.
28. Cole, Dame Margaret (1971). *The Life of G.D.H. Cole*, p. 51. Macmillan.
29. Martin, W. (1967). *The New Age under Orage*, pp. 123–5. Manchester University Press.
30. Orage, A. R. (1917). *An History of Economics*. Fisher Unwin.
31. Finlay, J. L. (1972). *Social Credit: The English Origins*, p. 83. Queen's University Press, McGill.
32. MacDiarmid, H. (1934). Representative Scots (2) The Builder. Major C. H. Douglas. In Grassic Gibbon, L. and MacDiarmid, H. *Scottish Scene, or The Intelligent Man's Guide to Albyn*, pp. 154–5. Jarrolds.
33. For a longer discussion of Douglas's thought, see Finlay, J. L. (1972). *Social Credit: The English Origins*, Chapter 5. Queen's University Press, McGill.
34. Werskey, G. (1978). *The Visible College*, pp. 30ff. Allen Lane.
35. Soddy, F. (1922). *Cartesian Economics. The Bearing of Physical Science Upon State Stewardship*. Henderson's.
36. For a recent and complete discussion of Soddy's economic thought see Daly, H. E. (1986). The Economic Thought of Frederick Soddy. In *Frederick Soddy (1877–1956): Early Pioneer in Radiochemistry*, (ed. G. B. Kauffman), pp. 119–218. Reidel, Dordrecht.
37. Mirowski, P. (1989). *More Heat than Light. Economics as Social Physics: Physics as Nature's Economics*, Chapter 5 passim. Cambridge University Press.
38. Soddy, F. (1922). *Cartesian Economics. The Bearing of Physical Science Upon State Stewardship*, p. 3. Henderson's.
39. Soddy, F. (1922). *Cartesian Economics. The Bearing of Physical Science Upon State Stewardship*, p. 9. Henderson's.
40. Ruskin, J. (1968). *Unto This Last*, pp. 179–80. Everyman Edition, Dent. In this form quoted by Soddy, F. (1922). *Cartesian Economics. The Bearing of Physical Science Upon State Stewardship*, p. 32. Henderson's.
41. Soddy, F. (1922). *Cartesian Economics. The Bearing of Physical Science Upon State Stewardship*, p. 32. Henderson's.
42. *Credit Power*, January 1924.
43. Soddy, F. (1924). *The Inversion of Science and a Scheme of Scientific Reformation*. Henderson's.
44. Soddy, F. (1924). *The Inversion of Science and a Scheme of Scientific Reformation*, p. 6. Henderson's.
45. See, for example, review of *Cartesian Economics*. *Public Welfare*, September 1922.
46. Soddy, F. (1924). *The Inversion of Science and a Scheme of Scientific Reformation*, p. 19. Henderson's.
47. Soddy, F. (1924). *The Inversion of Science and a Scheme of Scientific Reformation*, p. 28. Henderson's.
48. Soddy, F. (1924). *The Inversion of Science and a Scheme of Scientific Reformation*, p. 39. Henderson's.

49. Soddy, F. (1924). *The Inversion of Science and a Scheme of Scientific Reformation*, pp. 40–1. Henderson's.
50. Details and quotation from Finlay, J. L. (1972). *Social Credit: The English Origins*, pp. 122–3. Queen's University Press, McGill.
51. *Public Welfare*, 21 September 1921, 4. This was the first number of the publication in the 'Social Credit Series; *Credit Power*, November 1922.
52. This material comes from Finlay, J. L. (1972). *Social Credit: The English Origins*, Chapter 6. Queen's University Press, McGill.
53. Finlay, J. L. (1972). *Social Credit: The English Origins*, p. 124. Queen's University Press, McGill.
54. Finlay, J. L. (1972). *Social Credit: The English Origins*, p. 125. Queen's University Press, McGill.
55. *Public Welfare*, September 1922.
56. *Credit Power*, January 1924.
57. Reckitt Papers, 18/11.
58. Finlay, J. L. (1972). *Social Credit: The English Origins*, pp. 148–9. Queen's University Press, McGill.
59. Finlay, J. L. (1972). *Social Credit: The English Origins*, p. 153. Queen's University Press, McGill.
60. Finlay, J. L. (1972). *Social Credit: The English Origins*, pp. 155–6. Queen's University Press, McGill.
61. Finlay, J. L. (1972). *Social Credit: The English Origins*, pp. 157–63. Queen's University Press, McGill.
62. Finlay, J. L. (1972). *Social Credit: The English Origins*, p. 132. Queen's University Press, McGill. See also Levitas, R. (ed.) (1985). *The Ideology of New Right*. Polity, Cambridge.
63. Finlay, J. L. (1972). *Social Credit: The English Origins*, p. 128. Queen's University Press, McGill.
64. Thurlow, R. C. (1980). The Return of Jeremiah: the Rejected Knowledge of Sir Oswald Mosley in the 1930s. In *British Fascism: Essays on the Radical Right in Inter-War Britain*, (ed. K. Lunn and R. C. Thurlow), pp. 100–15. Croom Helm.
65. Thurlow, R. C. (1980). The Return of Jeremiah: the Rejected Knowledge of Sir Oswald Mosley in the 1930s. In *British Fascism: Essays on the Radical Right in Inter-War Britain*, (ed. K. Lunn and R. C. Thurlow), p. 101. Croom Helm.
66. Thurlow, R. C. (1980). The Return of Jeremiah: the Rejected Knowledge of Sir Oswald Mosley in the 1930s. In *British Fascism: Essays on the Radical Right in Inter-War Britain*, (ed. K. Lunn and R. C. Thurlow), p. 104. Croom Helm.
67. *New Britain*, 2 May 1934, p. 723.
68. Finlay, J. L. (1972). *Social Credit: The English Origins*, pp. 131–3. Queen's University Press, McGill.
69. Finlay, J. L. (1972). *Social Credit: The English Origins*, pp. 143, 186. Queen's University Press, McGill.
70. Soddy, F. (1926). *Wealth, Virtual Wealth and Debt: The Solution of the Economic Paradox*. George Allen and Unwin Ltd.
71. Soddy, F. (1926). *Wealth, Virtual Wealth and Debt: The Solution of the Economic Paradox*, Chapters 2 and 3. George Allen and Unwin Ltd.
72. Soddy, F. (1926). *Wealth, Virtual Wealth and Debt: The Solution of the Economic Paradox*, p. 10. George Allen and Unwin Ltd.

73. Soddy, F. (1926). *Wealth, Virtual Wealth and Debt: The Solution of the Economic Paradox*, p. 212. George Allen and Unwin Ltd.
74. Soddy, F. (1926). *Wealth, Virtual Wealth and Debt: The Solution of the Economic Paradox*, p. 303. George Allen and Unwin Ltd.
75. Soddy, F. (1926). *Wealth, Virtual Wealth and Debt: The Solution of the Economic Paradox*, p. 304. George Allen and Unwin Ltd.
76. Myers, M. G. (1940). *Monetary Proposals for Social Reform*, p. 71. Columbia University Press, New York.
77. Martinez-Alier, J. with Schlupmann, K. (1987). *Ecological Economics: Energy, Environment and Society*, p. 144. Basil Blackwell, Oxford.
78. Myers, M. G. (1940). *Monetary Proposals for Social Reform*, pp. 70–1. Columbia University Press, New York; Redman, T. (1991). *Ezra Pound and Italian Fascism*, p. 151. Cambridge University Press.
79. Walker, G. (1988). *Thomas Johnson*, p. 119. Manchester University Press.
80. Soddy, F. (1927). *The Wrecking of a Scientific Age*. Henderson's.
81. Soddy, F. (1927). *The Wrecking of a Scientific Age*, p. 6. Henderson's.
82. Soddy, F. (1927). *The Wrecking of a Scientific Age*, p.7. Henderson's.
83. Soddy, F. (1928). *The Impact of Science Upon and Old Civilisation*. Henderson's.
84. Soddy, F. (1928). *The Impact of Science Upon and Old Civilisation*, p. 21. Henderson's.
85. For example, on the return to the gold standard, see Chapter 17 of Moggridge, D. E. (1992). *Maynard Keynes: An Economist's Biography*. Routledge; Morris, M. et al. (1976). *The General Strike*. Penguin Books, Harmondsworth.
86. Taylor, A. J. P. (1965). *English History 1914–1945*, pp. 287, 291. Oxford University Press. This edition Penguin Books, Harmondsworth.
87. Blythe, R. (1964). *The Age of Illusion: England in the Twenties and Thirties 1919–1940*, p. 26. Penguin Books, Harmondsworth.
88. Jay, K. (1964). Introductory chapter to Gowing, M. *Britain and Atomic Energy 1939–1945*, p. 15. Macmillan.
89. Jay, K. (1964). Introductory chapter to Gowing, M. *Britain and Atomic Energy 1939–1945*, pp. 15–17. Macmillan; Badash, L. (1986). The Suicidal Success of Radiochemistry. In *Frederick Soddy (1877–1956): Early Pioneer in Radiochemistry*, (ed. G. B. Kauffman), pp. 27–41. Reidel, Dordrecht.
90. The Cassel Cyanide Company Ltd. Anonymous typescript in Bristol University Library.
91. Grassic Gibbon, L. and MacDiarmid, H. (1934). *Scottish Scene, or The Intelligent Man's Guide to Albyn*, p. 148. Jarrolds.
92. Strachey, J. (1928). *Socialist Review*, July 1928, p. 1.
93. Newman, M. (1989). *John Strachey*, pp. 19–20. Manchester University Press.
94. Soddy, F. (1928). *Socialist Review*, August 1928, pp. 28–30.

CHAPTER 7

New Britain

To understand Soddy's development during the next few years, we need to look at the New Britain Group, later known as the New Europe Group. This organisation is now relatively unknown but, during the nineteen thirties, it gained a considerable following proposing reforms based on a tripartite programme of cultural, economic and parliamentary changes which was developed through the period. The economic changes were closely based on Soddy's proposals and he was to become an influential member of the group. However, to understand the group during the thirties it is necessary to go back in time to the foundation of the group in inter-war London. This cultural world was characterised by numerous societies, clubs and associations, almost as many as there were individuals involved. Many of these groups had no formal membership, meeting place or publication, most of them were short-lived, while their agendas and programmes varied from merely eating and drinking to saving the whole world. Those who belonged to this world tended to associate with a range of organisations, passing from one to another according to what now look like whims, fancies and fashion. The favoured description for these organisations was 'group' and this will be adopted as the most suitably loose designation. Most of them were geographically near each other, in London and generally near or in Bloomsbury. Even the original 'Bloomsberries', although by the late 1920s having their centres in Sussex and France, maintained close links with this area, and with the other, newer groups. The attraction of this part of London was due in part to historical accident—it had long been associated with an intellectual society, and this had led to or depended on the number of cafes, tearooms, restaurants and pubs where informal meetings and conversations could be held. The nearby houses were also still cheap and large enough to provide suitable premises for lecture rooms and offices.

Membership of the Bloomsbury Group was restricted. 'The club was not exclusive but the terms for belonging to it were so ethically and culturally high that a comfortably small membership was guaranteed.'[1] However, the range of interests of this small club typified the society. Painting, novels, poetry, gardening, politics and economic theory were a part. And, if the club was exclusive, the individual members were not, and could be found in other, more welcoming surroundings.[2]

Orage's *New Age* and its contributors were the nucleus of one of these more open groups. Contributors, or would-be contributors to the journal, and their acquaintances, could meet in the Chancery Lane ABC, the Kardomah Cafe in Fleet Street and the Cafe Royal, as well as a variety of homes and studios.[3] Discussion of literary style, of recent exhibitions and plays, poetry readings and political discussion occupied these meetings where there was no agenda or entry qualification and the production of the journal was the *raison d'être* of the assembly. However, through loose associations like this, some sense of community was established, some notion of a group, with some kind of allegiance to each other, and some way of bringing in wider attachments. Paul Selver has described how, at one of these gatherings, he met Ezra Pound and was then invited to one of Pound's poetry evenings.[4] By these means, there would also always be a readily available constituency for new ventures.

While there is no sign of Soddy's involvement in any part of this world, and it seems very unlikely that he would have enjoyed the informality and diversity, one of the organisations with which he was to become closely connected had its roots here. This was the New Europe Group, an informal group which grew out of a number of initiatives set up by Dimitri Mitrinovic, a Serbian refugee who had come to England in 1914.[5] During his early years in England, Mitrinovic 'was looking for likely people who would be willing to commit themselves, with him, to the creation of a new age.' This would be based on 'a nucleus of individuals who, by their example and work, might act to transform social life.'[6] One of the first people to show any interest in this was Philip Mairet, a conscientious objector and later editor of *The New English Weekly*, but at the time disenchanted with his craft of stained and painted glass, and generally searching for some direction in his life. He and Helen Soden, a middle-class wife of a doctor, were Mitrinovic's first followers in England.

Mairet introduced Mitrinovic to Patrick Geddes, who was briefly in London in 1915. At this stage, Geddes was unimpressed, but was to renew contacts later. Immediately, Mairet's acquaintance led Mitrinovic in other directions. The first of these was to the Gill community, at the time in Ditchling.[7] Ethelmary Mairet, wife of Philip, was a weaver who joined Gill's community in 1918, shortly before Mairet resigned from the Red Cross. He settled in Ditchling, working as a farm labourer. Mitrinovic visited them there and it became something of a base for him. Helen Soden also settled in the village and after the war Mitrinovic rented Woodbine Cottage, a small house in Ditchling to which he retreated for periods of rest and recuperation.[8]

By 1920 Mitrinovic was based in London, correctly realising that if he wanted to make any mark, he needed to be near the centre of cultural and political life. Here, he was introduced to Orage and his circle, and became 'the predominant figure in Orage's world for two or three years, and possibly more'.[9] Mitrinovic, under the name M. M. Cosmoi, contributed a column on 'World Affairs' to *New Age* until Orage's departure for Paris in 1922. The prose style of this column, even after editing by Orage, was regarded by some as incomprehensible. For example, in September 1920, he wrote,[10]

> The world, we believe, has a divine dharma or purpose, defined by its nature, and unalterable by any effort of man . . . We hold, moreover, that this purpose has now been clearly manifested; and it can be summed up in the phrase, the functional organisation of the world as one. Looking at the problem before us in the light of this affirmation, our judgement of values must depend, as we have said, on their value in relation to this end.

The unedited Cosmoi was even less comprehensible, as the following passage, written after Orage had agreed to cease his rewriting, shows:

> The polarity and correlativeness of Great Britain and her Empire on the one side, and of Russia and the Slavonic world on the other, is wholly revealed in the opposite yet convergent tendencies of the British and Slavonic racial instinct.

Mairet felt that his was 'a very moderate sample' of Mitrinovic's 'cosmic commentaries', but believed that while 'Orage felt the strongest disapproval of the style', he had an equally strong 'faith in their significance'.[11]

This was the point at which Major Douglas had made *New Age* a mouthpiece for Social Credit, and so the journal was advocating two strategies for reform, Douglas's purely practical monetary theory and Mitrinovic's abstruse ideas. Whether these were seen by Orage or his readers as two parts of the same programme is unclear, although Mairet described Mitrinovic's objective as a 'scheme of politico-cosmic propaganda in which Douglasism was to be merely the foundation-stone of the economic sub-section'.[12]

Any real understanding of Mitrinovic's intentions is difficult since there was a constant central ambivalence in his plans for political reform. This is the contradiction between individual development and public activity. His ultimate aim was world federation, brought about by personal improvement and the institution of better relationships within and between individuals, households and communities. Mitrinovic described his ambitions to Philip Mairet early in 1918, in terms that are worth quoting at some length because they reveal his singular approach:[13]

> . . . we want to change these things: we want men and the world to be better. We form societies to do this, subscriptions ten shillings a year; we write books and make propaganda. We have done plenty of this now for a long time. And we find, do we not, that other people make societies with ten-shillings-a-year subscriptions

and quite different ideas and propaganda about changing our world? It is evident that the work cannot begin until everyone has better idea and thinks differently. But this we cannot do unless we feel differently: and that is not possible until we *become* different beings . . .

For this, it is necessary to forget who you are as a living, breathing individual, but to remember that 'you are a centre of the universal consciousness':

You cannot do it alone . . . A you and I may become a we-spiritually. And these two persons could become three: then they could incorporate others, indefinitely. When this shall be rightly and really begun it will grow into a power of understanding that will change the mind of the human world.

Mitrinovic spent much of the time between this conversation and his death in 1953 trying to keep the balance between these two approaches to reform. On the one hand was the individual development which he felt to be essential, on the other the development of wider, large-scale movements from the 'ten-shillings-a-year clubs' and societies which aimed at political and social change. In turn, many of the initiatives set up by Mitrinovic threatened to develop in this public dimension. In turn, he dissolved each one to return the emphasis to the individual.

One of the earliest of the associations set up by Mitrinovic came about through his contact with another circle, rather different from Orage and the *New Age*. Valerie Cooper was a musician who taught dance and eurhythmics with a studio in Fitzroy Street. She was enormously impressed by him and his teachings, 'that a person such as he could exist was a perpetually increasing wonder for me. No matter what subject I spoke of he, as it were, took me by the hand and led me along that path beyond the furthest horizon I could ever have dreamed of . . .'[14] She remained one of his supporters for the rest of her life and, through her, Mitrinovic met a number of people who otherwise would have remained strangers to him. The studio was a meeting place for members of the Bloomsbury Group, and for many other artists including the potter Bernard Leach, the conductor Edward Clark, Iris Tree, Matthew Smith and Augustus John. Valerie Cooper shared the studio with Lilian Slade, who also had a house in Temple Fortune Lane in Golders Green which she put at Mitrinovic's disposal.

Valerie Cooper introduced Mitrinovic to Alfred Adler while Adler was on a lecture tour in England in 1926 and the two men were mutually impressed. It was agreed that Mitrinovic should form the British Branch of the International Society for Individual Psychology which was accomplished in 1927. The first meeting was held in the studio, but premises were found at 55, Gower Street to provide a lecture room and library with a study for Mitrinovic on the ground floor. Here a programme of lectures and discussions was to continue, in a number of forms, throughout the inter-war period. The society was divided into a number of different sections or schools. These included Education, Sociology, Philosophy, Arts and Crafts, Music, Eurhythmics, and were open to those who were interested. There was also a medical group for medical practitioners, a men's group and a women's group. By the beginning of 1929 most of these groups had organised their own programmes of meetings, but throughout the existence of the society Mitrinovic was the driving force and he gave over fifty of their lectures. Nevertheless, splits occurred in the International Society, largely because of the attention paid to what were seen as extraneous subjects like sociology, which worsened when the 'Chandos Group' allied itself.[15]

The Chandos Group had been formed by Mitrinovic in May 1926, and named after the restaurant where they held their meetings. Their purpose was to meet regularly and to discuss the national political situation. The initial members had nearly all known each other in the days of Orage's editorship of *New Age*; Philip Mairet, Alan W. T. Symons and others had also been associated with the Adler Society, while most of them had been involved with Social Credit, but with the 'other side' from Soddy after the Swanwick split. W. Travers Symons, who had led the opposition to the Sheffield group's proposals, was at this time a partner in his father's firm of marine insurance brokers; he had been to New Zealand, lived with the anarchist colony at Whiteways in Gloucestershire and been to India. He was a member of the ILP and lectured to them about Social Credit; he was also a founder member of the Adler Society and combined this with an apparently incompatible interest in Social Credit in the journal *Purpose*, which he edited from 1929–40.[16] Maurice Reckitt was a rich Quaker Christian socialist who in 1915 had taken part in founding the National Guilds League and had been the secretary from 1919 to 1920 before becoming converted to Social Credit. He had then been a leading member of the Social Credit organisation throughout the late twenties and early thirties.[17] Philip Mairet had been associated with the Arts and Crafts movement at Chipping Campden under C. R. Ashbee; then, as we have seen, was connected with the Ditchling Community and worked for Patrick Geddes; he also wrote for *New Age* where he came into contact with Social Credit. Having established the group, Mitrinovic left it. In his absence others joined including Colonel J. H. Delahaye, an independent politician, while G. D. H. Cole, Lewis Mumford and T. S. Eliot were semi-regular attenders.

As well as discussions, the Chandos Group produced two books. The first was *Coal*, edited in 1927 by Alan Porter, a founder member of the group who left very soon after its foundation, perhaps because he had no belief in Social Credit. It was published by the Hogarth

Press, reflecting the connections between Mitrinovic's groups and the Bloomsbury Group. The title is somewhat misleading as the text is as much a general discussion sparked by the crisis in industry after the 1926 strike with only the final chapter providing specific suggestions about coal and its future. The National Guilds background of the writers is revealed throughout the book, with suggestions for local councils to organise industries and the 'right objective of the worker's control of industry',[18] and their Social Credit beliefs in their conclusion that[19]

> the problem is only susceptible of permanent solution by a change of financial policy that is, in a technical sense, revolutionary ... it consists of a change in the cost-accountancy of industry or service, whereby retail prices will be such that the aggregate of money paid out to those engaged in industry will always be sufficient to purchase the resulting products.

Most strikingly, their ideas on the limited supply of coal are so similar to those expressed by Soddy from the 1900s onwards that it is difficult to believe there had been no contact between them.[20]

> Coal is a capital store of energy. In using it we are living on capital. No new coal is formed in the place of the coal we use. There is no coal harvest. There is no manufacture of coal. When the coal reserves of the country are exhausted we shall have emptied our present chief treasury of power.

Even Soddy's concerns about pollution are recorded here, echoing his pre-war concern about the 'ham and eggs cloud' over Glasgow, 'the burning of raw coal creates a smoke pall over our modern cities that interferes gravely with the actinic rays of the sun.'[21] In a way this is not surprising. Less than two years before the book's publication many of its authors had, of course, been members, with Soddy, of Social Credit.

The other Chandos Group book, published in 1929, was more obviously a general discussion of Mitrinovic's plans for reordering society and in many ways the blueprint for the New Europe and New Britain Groups which were to be formed in the early 1930s. *Politics: A Discussion of Realities*[22] was 'prompted by' the General Election of 1929 when the similarities of the proposals of the three major parties 'made a critical analysis of underlying attitudes even more necessary', while the likelihood of there being no clear majority for any party led to an examination of the question of whether 'oppositions must always oppose rather than contribute'.[23] This leads the authors of the book to question the fundamental basis of British political structures. The book begins with a discussion of the nature of democracy which insists that the possibility only of voting for one of a number of more or less undesirable candidates leads to serious problems.[24]

> [It] involves for its fulfilment not an occasional appeal to 'the will of the

people' on issues held by authority to be appropriate ... but the continuous and spontaneous operation of choice, decision and creative effort on every plane of communal living ... It demands, moreover, a technique of representation sufficiently complex to correspond to the versatility of man's nature.

The 'main issues of home politics' are, it is argued, unemployment, the minimum wage, housing, the status of the trade unions and agriculture and rural life. The solution of the problems in each of these areas is, centrally, one of money supply and of production. No details are given here as to possible improvements in the system, but this was to become one of the central planks of the reforms proposed by the New Britain Group. World problems occupy another chapter, with the general solution lying in some kind of federation. Each of these areas was to be taken up in more detail during the thirties, but in many ways *Politics* is a direct forerunner of the core of New Britain's thinking.

The central part of the argument can be summed up in the proposal for a tripartite organisation, the three categories, economic, political and cultural being those on which the three chambers of the New Britain were to be based. 'The ultimate aim of politics is such organisation as will free men to fulfil their economic and cultural needs.' The other single most important issue was world peace. Since the party system had failed to achieve either of these, an alternative was needed.[25] This was to be found in the individual. 'We may vote, when the occasion arises, and exhort others to do so, since it is a duty to use any means that are at hand, but our real hope is in the individual, in whom alone is the secret of freewill.'[26] Individuals, identifiably of some special but unspecified kind, will be recruited to a 'New Party'. They will be known by a 'new initiative in present problems' at work, in a wider circle of cultural and social life, in foreign affairs through increasing international associations of work and culture. This could all be done[27]

under the necessity of voting upon the usual party lines. Under such limiting conditions those who feel themselves the 'new party' will have to make the casting of votes secondary to their real objective, which is to raise the level of political conscience as a whole and find means to impress candidates with the precise meaning of their support or opposition ... This is difficult and usually ineffectual for individuals, but it is high time for all 'new-party' men, who are by necessity individualistic thinkers, to seek each other out and work as far as possible together.

They could form local groups, above party politics, but then divided along conventional party lines. To achieve this, the 'new party' argues for a new type of individual, transcending democracy. 'The possibilities of politics depend finally upon those citizens, whoever they may be, who can rise to world-orientation and maintain it in the affairs of the common life.'[28]

It is clear that aspects of this thinking feed into what becomes, after 1931, British Fascism and so it is perhaps no coincidence that Oswald Mosley's political initiative in 1930 after he had left the Labour Party carried the same name—The New Party. Mosley, as we have already seen, had also been influenced by Douglas and his Social Credit economic theories, but his thoughts were even closer to Kitson, since like others who split with Douglas he disagreed with the A+B Theorem. During the mid-twenties he had had contacts with the *Age of Plenty*, and his economic ideas developed in parallel with those of Kitson and by the early thirties he believed 'we have passed from the economics of poverty to the economics of plenty'.[29]

In 1932 the Chandos Group supported Orage on his return from America. Brenton, the editor by then, refused to give up his role on *New Age* in favour of Orage, who then started *New English Weekly* in the same offices with many of the same contributors. At this point, Orage returned to the political and economic journalism of his *New Age* days, in line with the thinking of the Chandos Group and also coinciding with the developing interests of one of the past contributors to *New Age*, Ezra Pound. Pound had by this time passed beyond Douglas's Social Credit, which he had discovered during the twenties, and had begun to write on economics. He had read and was impressed by Silvio Gesell, but was open to other ideas. These other ideas were significantly those of Soddy.[30] Through association with Pound and with Odon Por the new journal came to support Italy's corporate state while from others, like Horsfall Carter and Rolf Gardiner, came the desire to promote German fascism. Although there was a change of direction by 1933 there were still some regrets and a lingering belief that Hitler would support Social Credit.[31]

While in all these cases some version of fascism was seen to be strongest in the economic sphere, for Mitrinovic, 'fascism's concern with hierarchy and superiority of some people compared to others was equally correct when applied to the cultural sphere', while 'communism's emphasis on socialism and equality was absolutely correct when applied to the economic sphere.'[32] Although retaining the notion of the importance of the individual and of some kind of natural superiority, Mitrinovic and the New Britain and New Europe Groups were at pains to disassociate themselves from both fascism and communism. Their constant argument was that they were looking for a third way.

By 1930 the national and international situation had got very much worse. The Wall Street Crash of 29 October 1929 provoked a major crisis throughout the Western world. As prices fell, the producer countries suffered dramatically without the possibility of manufacturing being able to make up the deficits. Although to many contemporaries it was unemployment that was the central problem, to monetary theorists, the actual nature of the 'crash' pointed to the failure of the banking and market system which they had so long predicted.

Mitrinovic considered the circumstances to be so threatening that he decided that action beyond individual guidance and teaching was needed. The worsening crisis and the final collapse of the Labour Government in the summer of 1931 added to the sense of urgency. By 1930 the Adler Society had almost collapsed and it finally disintegrated in 1931. Partly as a replacement, and partly to reflect the new urgency in political and economic spheres, Mitrinovic launched the 'Eleventh Hour Flying Clubs', clubs only in the loosest possible terms, with nothing to do with flying, at least in the air. The title reflects the sense of urgency that was felt, and the need for speedy and effective action. This was explained by Helen Soden and Ethel Mairet, 'on behalf of women and the Ditchling Group', in a letter to Patrick Geddes. They wrote: 'in view of the extreme collapse and crisis in affairs at the present moment, extreme measures are forced upon us.'[33] This letter goes on to set out the scheme of the Eleventh Hour Flying Clubs,

... already inaugurated in Ditchling ... [they will] become a means of bringing about a re-inforcement of opinion and contact among people, so that a new living force will arise which will be strong and capable of influencing a new construction of [sic] regime for England and over Europe.

They were also going to ask other patrons to join, 'such as Professor Soddy'.

The reappearance of a conjuncture of Geddes and Soddy here is interesting. Neither had obviously had any connection with the Adler Society or the Chandos Group, and indeed they may have been antagonistic to aspects of them. Geddes had left England for India after his disagreeable first meeting with Mitrinovic in 1915, and Soddy had split from the Social Credit Movement after the Swanwick Conference in 1925 although he retained, as we have seen, many links with it. However, Geddes had remained in touch with various British developments. He had supported the Kibbo Kift Kin and had reviewed Hargraves's *Confession* in the *Sociological Review* in 1928. The publication of Amelia Defries's *The Interpreter Geddes* would have reminded the literary world of his existence even if, more generally, he felt that 'few are really interested'.³⁴ By 1930 Geddes was living in his Scots college in Montpelier, but taking trips to London during the summer to avoid the heat.³⁵

Soddy's connections were closer. Although after the Swanwick split he had joined the minority Social Credit branch, the Economic Freedom League, the gap between them and the Douglasite majority was never unbridgeable. By 1930 the Economic Freedom League had collapsed and Kitson had formed the Monetary Reform Association,³⁶ and while there is no evidence, it seems likely that Soddy went with him.

By this time, the worsening national and international situation was reflected in revitalised interest in schemes for economic reform and renewal. This is shown clearly in the creation by the Labour Government of the Economic Advisory Council in January 1930. Although it lasted only until the next year, its standing committee, the Committee on Economic Information, which included G. D. H. Cole and J. M. Keynes amongst its members, remained in business until it was incorporated into other wartime bodies.³⁷ In this atmosphere, attention was again paid to Social Credit and other monetary theories as a possible way out. Soddy's next book was published in this climate in 1931. *Money Versus Man* was one of 'The Library of New Ideas . . . provocative books upon widely discussed topics of the day', and was a more popular formulation of *Wealth, Virtual Wealth and Debt*.³⁸ In this work Soddy continued to advocate a form of socialism, but of a rather odd kind. He still aimed at the 'national ownership of the means of production, distribution and exchange' but this was to be effected 'not by oratory or by violence but by paying for it'.³⁹ Under his system of economics, wealth would be redistributed since production would increase, and so therefore would employment and the quantity of goods purchased. Taxation on these goods would provide a national income to provide for the unemployed. Eventually, scientific discovery would replace most kinds of hard labour, and some kind of national dividend, on the Douglas model, would replace wages. However, this would remain a dual system. The productive units would be privately owned, the state would take care of the monetary aspects.⁴⁰

The Eleventh Hour Clubs lasted a very short time, however. Groups had been set up and a journal, *The Eleventh Hour*, published. In the autumn of 1932 Arthur Kitson was their president, and they had organised a lecture series and a membership application which defined their object: 'To reform the money system of Great Britain by taking the power to create money credits out of private hands, and restoring it to the Crown and People', and their method was⁴¹

... by every means of propaganda and action, by speech, writings and publications of every kind, by forming branches throughout the country, co-operating and co-ordinating with other interested organisations and through *non-party* political action when deemed advisable.

The connection here with the Chandos Group plans for a new party is very obvious.

Despite these brave plans, by early 1933 the Eleventh Hour Flying Clubs had disappeared and their interests included in the New Europe Group (NEG). The relationship between the Eleventh Hour and the NEG is distinctly blurred. In the letter to Geddes from Ethel Mairet and Helen Soden, although they say that they are writing on behalf of the Eleventh Hour Flying Clubs, they 'are hoping to send you a delegation from the new European Group to lay before you proposals and plans for European Alliance . . . also the constitution of our new European Group'.[42] The group at Ditchling never seems to have actually formalised either its presidency or patrons for the Eleventh Hour Groups before the organisation changed its name to the NEG, of which Geddes was the first president. The NEG had offices in Gower Street where Winifred Gordon Fraser acted as secretary. However, it was the all-pervasive Mitrinovic, who lived on the premises, who dominated it for many years.

The organisation of the Gower Street premises at about this time was described by Willa and Edwin Muir who were 'disappointed and distressed' by what they saw of this world of the NEG.[43]

We were received by an incisive secretary who said that 'The Master' would appear in a little while . . . The room was fluttering with devotees also waiting to see the Master, and some of them talked to us. Apparently they were all laying their possessions at Mitrinovic's feet . . . When Mitrinovic came in, the whole room thrilled and moved reverently towards him . . . his mouth looked much the same, but his eyes did not, and his voice was portentous with self-importance. We non-joined his bogus cult . . . Some part of London, it appeared, was indeed Looney-bin.

In December 1931 the NEG organised the first of many series of lectures at 55, Gower Street, beginning with Mitrinovic on 'A United Europe in a World Order'. Speakers in this first series included Raymond Postgate, Ellen Wilkinson and Arthur Kitson, and the lectures were chaired by John Strachey, amongst others. A further series was organised at the Caxton Hall during the summer of 1932 and reflect the range of interests and areas of reform to be addressed by the NEG. They are also a remarkable testimony to the breadth of largely left wing opinion on which the NEG was able to draw. Under the general title of 'Popular Myths Exploded', Arthur Kitson demolished the notion 'That the Bank of England is Essential to Industrial Prosperity' and 'That the Gold Standard is Essential to Financial Stability' in meetings chaired by Oliver Baldwin. Soddy, with H. W. Nevinson in the chair, spoke on 'That Poverty is God'; Professor J. MacMurray entitled his contribution 'That Science Will See Us Through' in a meeting chaired by Leonard Woolf while C. E. M. Joad, with Harold Nicholson as chairman, argued 'That Civilisation is Civilised'. Raymond Postgate exploded the idea 'That Nationalism and Imperialism are Conducive to Human Welfare' and Ellen Wilkinson spoke on the proposition 'That Capitalism has Anything Further to Offer Us' and the ubiquitous John Strachey chaired a meeting.[44] The final 'myth', however, returned to Mitrinovic's long-term concerns, 'the separateness of men', in which he explored the significance of Europe in the evolution of the world. Since it was in Europe that individualism had reached its furthest point, 'Europe must take the initiative to turn individualism into communal recognition of personality and personal acceptance of community' and Britain should pioneer the route. 'Then will arise a New Britain which by her sound sense and courageous action will lead to World Socialism and World Peace.'[45]

During this period Soddy was at the height of his popularity and influence in areas outside science. Not only was his latest book republished and his lectures well-attended, but Dennis Wellesley Maxwell dedicated his book *The Principal Causes of Unemployment* to 'Professor Frederick Soddy MA FRS' in 1932, and G. D. H. Cole in 1933 edited *What Everybody Wants to Know About Money* which included a chapter on 'Frederick Soddy and the 100-Percent Reserve Plan' by H. T. N. Gaitskell.[46] Perhaps more importantly, though, he had at last found a group of congenial spirits who tolerated his unpredictable behaviour and outbursts of rage and who genuinely believed that what he was saying was important and useful. He was to remain associated with this group, through its various manifestations, through the rest of his life. From 1933 Soddy was president of the NEG, an office he held until his death in 1956. He followed two of his heroes in this office, Patrick Geddes (who died in 1932) and Arthur Kitson.

After the Caxton Hall lectures, and no doubt encouraged by their popularity, the first of many publications by the NEG was produced, the *New Britain Quarterly* with a first issue in October 1932. This magazine was unusual in size and format. It was 16 by 14 inches, laid out in three columns. Its content included articles by familiar names like Soddy and Mairet, but more unusually Mitrinovic 'decided to promote certain members of the group by printing under their names extracts of articles or lectures that had been produced by other group members'.[47] Mitrinovic himself never wrote in *New Britain Quarterly* although his control can be seen throughout. The first issue sold 2000 copies, perhaps because of its unusual format, but very quickly declined. It lasted for only three issues under this title and then, in 1933, two periodicals existed for a while. The large format *New Britain* was re-titled *The New Atlantis For Western Renaissance and World Socialism*, while on 24 May 1933 *New Britain* became a weekly newspaper, 'the sixpenny weekly for two pence',[48] but retained its links with the past with a regular cartoon and slogan 'The Age of Plenty'. *New Atlantis* existed for only two issues before transmogrifying into *New Albion For British Renaissance and Western Alliance* for one edition, after which it reverted to *New Britain For British Revolution and the Social State* in Autumn 1934, as a result of merging *New Albion* with *New Britain Weekly*. The weekly publication was henceforth to be called *Eleventh Hour For New Britain*.

While the titles are dizzying, the purpose and content of these publications varied little. All the publications were directed to the aims of monetary reform, of world federation and of individual development. The contributors to them all were very similar although, as with the lectures, remarkably mainstream and very impressive. Among the regular contributors were Hugh McDiarmid, the Scottish poet; J. T. Murphy, trade unionist and one of the founders of the British Communist Party; Jack Common, proletarian novelist and friend of George Orwell; Basil Boothroyd, later a Conservative MP, and John Grierson, the pioneer film maker. While Soddy contributed to almost every issue of the monthly, and is mentioned in the weekly, his articles in the weekly are rare. The reason for this is not really clear, although there is a sense that Soddy had some disagreement with Purdom, the editor of the weekly, since he contributed only to the very early issues and then the last ones, after Purdom's resignation. This is possibly also because he did not really approve of the popular version; he felt that 'New Britain was altogether too much an ordinary interesting weekly to be the organ of the movement.'[49]

There were some other, more noticeable differences between the monthly and the weekly publications. Most obviously, there was the price, two shillings and sixpence for the monthly, two pence for the weekly. The monthly, no matter what its title, was most obviously under the control of Mitrinovic. It was a glossy paper sometimes with coloured illustrations from recent exhibitions or in one instance, of Indian art. The articles tended to be longer and more theoretical than those in the weekly. The monthly, as we have seen, constantly changed its name while the weekly continued with one title for the two years of its existence.

The reason for the constant changes of name is unclear. The name never actually affected the style or the content or the paper, or even the design of the cover, it was only the name that changed. It may have been a game played by Mitrinovic as a part of his policy of keeping his followers alert and avoiding any sense of false security. It may equally well have been an attempt to keep up sales by an appeal to novelty.

The weekly was, from the beginning, designed to appeal to a larger and more popular audience. From 1933, the *New Britain* weekly grew in popularity. By August 1933, after three months' publication, sales amounted to 32 000. It appealed to a wide section of the population. There were contributions from Harold Macmillan, Bertrand Russell, Odon Por, J. T. Murphy, C. M. Grieve (Hugh McDiarmid), T. S. Eliot, articles on political and monetary reform, reviews of the arts, short stories, poems and reproductions of a range of works of art and photographs of articles and buildings of cultural and architectural importance. Each issue contained a call for support—for money, increased sales and, most significantly, for the New Britain Groups. By August 1933 sixty New Britain Groups had been formed. The aims of these Groups were very similar to those of the NEG, but were more clearly stated in almost every issue of the magazine. They were:

1. The belief that an altogether new and different Britain is necessary and possible.

2. The conviction that in this emergency the initiative of every British man and woman is called for.

3. The affirmation that the perfection of the individual is the true aim of national existence.

4. The personal alliance of all who believe that Britain should be transformed into a Social State.

5. The immediate and thorough adaptation of production and distribution to realize the new age of plenty.

6. The guidance of the national wealth processes by the direct producers.

7. The federation of European nations leading to and forming the basis of world federation.

In more detailed manifestos these proposals were elaborated and their roots in social credit, national guilds and Mitrinovic's proposed reforms can be seen.[50]

1. The complete overhaul of the monetary system by restoring the right to issue credit to the nation rather than the banks.

2. The re-organisation of industry as National Guilds based on worker's control.

3. The devolution of parliament into three Chambers—a House of Industry based on the National Guilds taking control over economic affairs,

a House of Culture composed of representatives of the arts, sciences, religion and education which would exercise a guiding influence over cultural affairs, and a Political Chamber which would be concerned with questions of law and order and international relations.

4. The utmost devolution of decision-making power on as wide a range of issues as possible within Britain itself as a step on the way towards European and ultimately world federation.

New Britain Groups were formed through contacts in the journal, which found its audience in a number of ways. It was sold at bookstalls, on the streets by members of the groups and was, at least in Oxford, left in common rooms for undergraduates to read. One undergraduate who discovered the organisation in this way spoke of his excitement at his discovery. He had tried political parties while an Oxford undergraduate and decided that the only possible one was the Communist Party but felt that the communists were 'reliant on slogans' about revolution. He found a copy of *Eleventh Hour* and through it discovered New Britain and Mitrinovic and decided that this was the organisation for change and a peaceful Europe. He communicated his feelings to at least one old schoolfriend who also joined,[51] and then sold the *Eleventh Hour* on the streets with Maitland, now Lord Lauderdale. His membership continued after graduation when he lived on his National Savings of £100 for a couple of years so that he could study with Mitrinovic. When his money ran out he worked in the City, but continued his membership until his recent death.[52] The schoolfriend also joined and remains one of the New Atlantis Group.

Through enthusiasm like this, Oxford developed both a town and a gown branch of New Britain, with Soddy as the '"don" responsible' who 'signed the proctorial form of application for permission to join New Britain from an undergraduate society at Oxford'.[53]

A slightly different way of finding the Group is described by one of the women members. After a conventional middle-class education at a Quaker boarding school with 'Clarkes, Rowntree and Cadbury daughters' where she learnt the tradition of good works she took a diploma in social work and geography at Leeds University, wanting to become hospital almoner. She was shocked by the poverty in Leeds slums and so determined to find a way of 'changing society'. After moving to London, she attended a New Britain lecture series against war, and like many others, felt she had found a way of helping to improve society and of avoiding war. She 'joined the organisation', met Valerie Cooper and another woman from her Studio and in turn joined Cooper's School of Human Movement in Fitzroy Square. Finding somewhere to live opposite, Cooper became her mother/sister figure. For much of this period she lived on a £5 a week allowance from her father.[54]

Money on which to live was a difficulty for the members of the inner circle of New Britain. The demands on their time made full-time work an immense strain, but the organisation had no funds on which they could draw for personal expenses. As a result, willing recruits had to be rejected. Soddy described to Winifred Gordon Fraser his correspondence with one such potential recruit who wanted 'paid employment by the New Britain or other organisation. I told him . . . you can do what you do because people gave their services voluntarily so freely. I told him also I did not advise anyone not to some extent economically independent to go in for this sort of work at all . . .'[55]

Looking at the programme put forward by New Britain, it is clear that its origins lie in the work of the Chandos Group in the 1920s. The parliament devolved into three chambers and the 'complete overhaul of the monetary system' come directly from the Chandos Group. Other aspects of the programme show a mixture of intellectual origins which are remarkably like those of Soddy himself, tracing a journey through the world of monetary reformers. Here we see aspects of the National Guilds, Social Credit and even leftist socialism. Nevertheless, there are disturbing sides to the programme. Its celebratory tone has much of the feel of European fascism, while its open corporatism looks in the same direction. It is significant that despite the general tone of *New Britain*, as we said earlier, being left wing, it did publish pro-fascist or crypto-fascist articles, like those on the corporate state by the Italian Odon Por, and the occasional contribution by Major-General J. F. C. Fuller.[56]

While this programme was attractive to many, it contained no sense of what could be done. The nearest to a programme for action was given in Mitrinovic's inimitable style in 'New Britain Manifestos' in the *New Britain Quarterly* as[57]

To wait for leaders is to evade responsibilities.

When the people decide what they want there will arise among them men and women through whom they can speak. The time has come for those to come together who know what sort of a life they want and how they want it to be organised.

Those who wish to save themselves from drifting into a state of war—a war of all against all, must make themselves responsible to each for all, and find others who will join them in overcoming all that stands in the way of a NEW ORDER.

Even here there is little indication of a programme for action since, in so far as Mitrinovic had an agenda, it was disguised as almost a secret society, the key role of which was to develop the individual and thus to provide the leaders in the New Britain which was to come.

From the very beginning of his activities after the war, when his only pupils/followers were Philip Mairet and Helen Soden, Mitrinovic had exaggerated his role as the 'mysterious slav', although he seems also to have had a particularly powerful presence. We have already quoted Valerie Cooper's powerful first impressions, but she was not alone. Philip Mairet wrote that after their first meeting 'my imagination was already investing him with the aura of sage or prophet',[58] while Paul Selver felt his power even more strongly, 'hardly had I shaken hands with him than I found myself so affected by his mere presence that I nearly lost consciousness.'[59] Even those who were more cynical about him generally agreed about his attraction. To Willa Muir he was 'a source of joy' even if he was also 'an egregious nonsense-monger'[60] and Alan Watts 'loved and feared him, for my Buddhist and Theosophical friends were of the opinion that he was a black magician'.[61] He used this power, deliberately or unconsciously, to create a close group, an inner circle, around him with whom he developed what one of his followers called 'the method of "personal alliance" . . . a no-holds-barred psychoanalysis [in which] for some months, we resolutely destroyed and rebuilt each other's personalities.'[62] While this group was more 'private' than the political initiatives and public letters, there was an even more particular group, closer to Mitrinovic and open only to those personally invited. Alan Watts was one of those invited and described his 'joining' like this. It took place in Mitrinovic's 'sanctum sanctorum' where he lived semi-secretly, 'surrounded by devoted disciples and adoring women'.[63] Here,[64]

I found him [Mitrinovic] sitting at the head of his bed like plump Buddha . . . he said, 'Alan Watts, I love you but I do not like you. Nevertheless, I am going to invite you to join an eternal and secret fellowship which will watch you, guard you, and keep track of you wherever you may go in the world. We call it the Wild Woodbines, named after the cheapest cigarettes in England. Every member is to carry a package, and the sign of recognition is to produce your package and say "Have one of mine" . . .'

If not at the time, Alan Watts was afterwards deeply cynical about Mitrinovic and his actions on this personal level. Although believing that the changes advocated by New Britain might have worked, he suggested that 'with the institution of the Woodbines, Mitrinovic effectively went out of politics' and certainly it seems to have done so for him. Writing forty years after the events, he remembered the 'Woodbines' as concentrating on intimate *carpe diem* parties' and as a meeting place where he met several of his loves of the period.[65]

However, there is an alternative interpretation of this inner circle, the one put forward by those who remained part of the circle. This was that the chosen few would be the foundation of a Senate, the leaders of the New Britain. The first public task of the inner group was to destroy the growing plans for a wider political body. By November 1933 pressure for such a body had grown to such a point that C. B. Purdom, the editor of *New Britain* weekly, wrote in an editorial:[66]

A New Britain Movement needs to exist. Until a movement is in being with declared aims, a defined policy, and a programme of action, the proposals we discuss in these pages must remain vague . . .

Although directly opposed to the kind of progress suggested by Mitrinovic, this proposal was attractive to a number of London and provincial Groups and a split between the two approaches seemed inevitable. The London Group organised a series of conferences at which a 'Draft Constitution' was produced and a suggestion for Regional Councils developed. By the end of December 1933, a London Regional Council was brought into being as a first stage in the '"Devolution", practised as well as preached', one of the principles of New Britain. By 3 January 1934, there was a new London office in 3, Gordon Square, separate from the National Organisation, still at 60, Gower Street. The differences between these groups were to be discussed and possibly voted on at the National Conference at Leamington Spa in March 1934. However, the conference became the site of a tactical battle between the two sides. Mitrinovic described himself as Bakunin rather than Marx. He was for anarchist rather than democratic processes and was against voting as a demonstration of the will of the meeting. Purdom and the London Groups were for democratic processes, they wanted votes, a clear organisation at each level, each reporting to the one above and some kind of elected leaders. All this was in many ways reminiscent of the Social Credit conference at Swanwick where a national organisation had been proposed. Much the same mechanism was to be put into operation at Leamington Spa. As at Swanwick, an additional meeting to those on the programme was announced, this time tacked on to the end of Soddy's Saturday evening lecture. After he had stopped speaking, Lilian Slade, one of Mitrinovic's strongest supporters, proposed 'That this conference should solve the problems of leadership which must arise in the early stages of a movement by appointing six of those here who obviously had the confidence of the delegates. And one other who was known to many.' Her proposal was accepted by general acclaim, and then, when the rest of the conference discovered the coup on the Sunday, it was put to the vote and the result was only two oppositions out of one hundred and twenty seven to the proposed election. Soddy was one of the six. Unlike at Swanwick, he had this time found himself on the winning side.

It is impossible to know with any certainty why Soddy should have taken the opposite position to the one he had taken in 1925 but it is possible to suggest some reasons. It seems likely that he had lost faith in democratic mass movement organisations. This can be the only reason for his accepting the role as one of the 'leaders'. This could well have been reinforced by a sense that such movements had anyway failed conspicuously in his terms. More importantly, perhaps, he had, within the New Britain and New Europe Groups, identified himself with the 'official group', rather than the new London Alliance. We have seen how he was more closely allied with the monthly, Mitrinovic dominated paper, rather than the weekly. Beyond this, Winifred Gordon Fraser, on behalf of the newly elected leaders, argued that 'we can now proceed to organise ourselves much more effectively now that the movement is united'.[67] Despite these arguments, Soddy's total commitment to this new organisation is open to question since, immediately after the conference, just as the new leaders set about speaking to the groups to convince them that their election was a good thing, even if not totally constitutional, Soddy went on holiday to Buttermere in the Lake District.[68]

However, this was also the beginning of the end of any national, influential party. Purdom resigned as editor of *New Britain*. The quality of the journal declined and so did its readership. There was an outcry amongst those who had joined to prevent war over Mitrinovic's suggestion that Europe should rearm to prevent Hitler taking power, and this led to regional groups severing their connections. The patrons of the journal were reluctant to continue paying and so funds dried up, and often the printer's bill could be paid only after 'money runs'. Members of the group closest to Mitrinovic were sent with lists of names to beg money for each issue. By August 1934 these methods finally failed and the last issue of the *New Britain Weekly* was published. Although different initiatives were attempted over the next couple of years, New Britain as a movement was over.

However, the New Europe Group, as an inner circle around Mitrinovic, continued. He retained the notion of a select band who would be prepared to take power in the future and who could be trained for their future tasks through participation in a number of different groups. These groups took a number of forms. They were based on 'natural differences': male and female, optimist and pessimist, and on other lines of Mitrinovic's arranging. The members of these groups would form the Senate. This was not 'a distinct body separate from all others, but a large and loosely connected group of persons with members in every other function of the community . . . Its function would not be leadership, as generally understood, or government, but that of relating all other functions, economic, cultural and political, to one another . . .'[69] However, there were still different degrees in this system. There was the central group, living with or near Mitrinovic, sharing various household tasks and expenses as well as the kind of psychological experiments described by Alan Watts. At the other extreme there were the 'VIPs', the patrons or 'important personages' who were useful in a number of ways. They did not share the group experiences, but were always welcome to the households, where they were treated as very special guests. Soddy was of this number, together with Major-General J. F. C. Fuller, S. G. Hobson, Ben Tillett, and Charles Purdom. There were then various other groups in between.

It has been suggested that this kind of arrangement was what Mitrinovic had aimed at, that he wanted personal power over a relatively small group, not a national movement in which he would play a relatively small part, and that the national movement had been the result of him losing control in some way. This is possible. Certainly no further attempt was made to form another large-scale movement. But this could have been the effect of the national and international situation. While the New Britain Movement had been growing and then destroying itself, Europe had been moving nearer and nearer to open conflict. As the links between monetary reform and fascism were close, perhaps the dangers of a political movement at this stage were considered too great.

Whatever the cause, the New Europe Group continued in its small, exclusive form, changing its name to the Renaissance Club in 1946 and finally the New Atlantis Foundation from 1954 and still functioning today. They moved from Bloomsbury to Richmond and then to Ditchling where they managed to buy Woodbine Cottage, where Mitrinovic had lived, as the centre of their activities. Throughout this period they organised lectures and published pamphlets on what they considered important issues of the day. The latter activities of the group will be considered in the following chapter, together with other, later monetary reform groups with which Soddy had connections.

Notes and references

1. Blythe, R. (1964). *The Age of Illusion: England in the Twenties and Thirties 1919–1940*, p. 123. Penguin Books, Harmondsworth.
2. For example, Leonard Woolf and Harold Nicholson chaired lectures organised by the New Europe Group in 1932. New Atlantis Archive.
3. Martin, W. (1967). *The New Age under Orage*, p. 45. Manchester University Press.
4. Selver, P. (1959). *Orage and the New Age Circle*, pp. 34–6. George Allen and Unwin Ltd.
5. Unless otherwise noted, information about Mitrinovic and his various initiatives is from Rigby, A. (1984). *Initiation and Initiative: An Exploration of the Life and Ideas of Dimitrije Mitrinovic*. Columbia University Press, New York. The spelling of 'Dimitri' varies in different works so, except where specifically quoted, 'Dimitri' is the form used.

6. Rigby, A. (1984). *Initiation and Initiative: An Exploration of the Life and Ideas of Dimitrije Mitrinovic*, p. 50. Columbia University Press, New York.
7. MacCarthy, F. (1989). *Eric Gill*, Chapter 8. Faber and Faber.
8. Mairet, P. (1981). *Autobiographical and Other Papers*, (ed. C. H. Sisson), p. 128. Carcanet, Manchester.
9. Mairet, P. (1981). *Autobiographical and Other Papers*, (ed. C. H. Sisson), p. 178. Carcanet, Manchester.
10. This is an extract, chosen for its simplicity, quoted in *Certainly, Future: Selected Writings by Dimitrije Mitrinovic*, (ed. H. C. Rutherford), p. 87. Columbia University Press.
11. Mairet, P. (1936). *A.R. Orage: A Memoir*, p. 83. J. M. Dent and Sons Ltd.
12. Mairet, P. (1981). *Autobiographical and Other Papers*, (ed. C. H. Sisson), p. 83. Carcanet, Manchester.
13. Mairet, P. (1981). *Autobiographical and Other Papers*, (ed. C. H. Sisson), pp. 103-4. Carcanet, Manchester.
14. Quoted in Rigby, A. (1984). *Initiation and Initiative: An Exploration of the Life and Ideas of Dimitrije Mitrinovic*, p. 63. Columbia University Press, New York.
15. Finlay, J. L. (1972). *Social Credit: The English Origins*, pp. 95, 182. Queen's University Press, McGill.
16. Peart-Binns, J. S. (1988). *Maurice Reckitt: A Life*, p. 103. The Bowerdean Press and Marshall Pickering, Basingstoke.
17. Reckitt papers, box 18; Finaly, J. L. (1972). *Social Credit: The English Origins*, p. 120. Queen's University Press, McGill.
18. Demant, V. A. et al. (1927). *Coal: a Challenge to the National Conscience*, p. 42. Hogarth Press.
19. Demant, V. A. et al. (1927). *Coal: a Challenge to the National Conscience*, p. 66-7. Hogarth Press.
20. Demant, V. A. et al. (1927). *Coal: a Challenge to the National Conscience*, p. 56. Hogarth Press.
21. Demant, V. A. et al. (1927). *Coal: a Challenge to the National Conscience*, passim. Hogarth Press.
22. *Politics: A Discussion of Realities* (1929). Initiated by J. V. Delahaye in company with H. Cousens, V. A. Demant, Philippe [sic] Mairet, A. Newsome, A. Porter, M. B. Reckitt, W. T. Symons. The C. W. Daniel Company.
23. *Politics: A Discussion of Realities* (1929). Initiated by J. V. Delahaye in company with H. Cousens, V. A. Demant, Philippe [sic] Mairet, A. Newsome, A. Porter, M. B. Reckitt, W. T. Symons. pp. 10-11. The C. W. Daniel Company.
24. *Politics: A Discussion of Realities* (1929). Initiated by J. V. Delahaye in company with H. Cousens, V. A. Demant, Philippe [sic] Mairet, A. Newsome, A. Porter, M. B. Reckitt, W. T. Symons. p. 55. The C. W. Daniel Company.
25. *Politics: A Discussion of Realities* (1929). Initiated by J. V. Delahaye in company with H. Cousens, V. A. Demant, Philippe [sic] Mairet, A. Newsome, A. Porter, M. B. Reckitt, W. T. Symons. p. 32. The C. W. Daniel Company.
26. *Politics: A Discussion of Realities* (1929). Initiated by J. V. Delahaye in company with H. Cousens, V. A. Demant, Philippe [sic] Mairet, A. Newsome, A. Porter, M. B. Reckitt, W. T. Symons. p. 169. The C. W. Daniel Company.
27. *Politics: A Discussion of Realities* (1929). Initiated by J. V. Delahaye in company with H. Cousens, V. A. Demant, Philippe [sic] Mairet, A. Newsome, A. Porter, M. B. Reckitt, W. T. Symons. p. 172. The C. W. Daniel Company.

28. *Politics: A Discussion of Realities* (1929). Initiated by J. V. Delahaye in company with H. Cousens, V. A. Demant, Philippe [sic] Mairet, A. Newsome, A. Porter, M. B. Reckitt, W. T. Symons. p. 179. The C. W. Daniel Company.
29. Thurlow, R. C. (1980). The Return of Jeremiah: the Rejected Knowledge of Sir Oswald Mosley in the 1930s. In *British Fascism: Essays on the Radical Right in Inter-War Britain*, (ed. K. Lunn and R. C. Thurlow), p. 104. Croom Helm.
30. Redman, T. (1991). *Ezra Pound and Italian Fascism*, p. 99. Cambridge University Press.
31. Finlay, J.L. *Social Credit: The English Origins*, pp. 130, 174–7. Queen's University Press, McGill.
32. Rigby, A. (1984). *Initiation and Initiative: An Exploration of the Life and Ideas of Dimitrije Mitrinovic*, p. 114. Columbia University Press, New York.
33. Letter to Geddes from 'Women and the Ditchling Group, Ethel Mairet and Helen Soden'. October 1931. SUA T-Ged 9/1841.
34. Boardman, P. (1978). *The Worlds of Patrick Geddes*, p. 387. Routledge and Keegan Paul.
35. Biography of Patrick Geddes by Abbie Ziffren. Part 1 of Stalley, M. (1972). *Patrick Geddes: Spokesman for Man and the Environment*, p. 99. Rutgers University Press, New Brunswick.
36. Finlay, J. L. (1972). *Social Credit: The English Origins*, p. 128. Queen's University Press, McGill.
37. Winch, D. (1969). *Economics and Policy*, p. 265. Hodder and Stoughton.
38. Soddy, F. (1931). *Money versus Man: A Statement of the World Problem from the Standpoint of the New Economics*, back cover. Elkin Mathews and Marrot.
39. Soddy, F. (1931). *Money versus Man: A Statement of the World Problem from the Standpoint of the New Economics*, p. 103. Elkin Mathews and Marrot.
40. Soddy, F. (1931). *Money versus Man: A Statement of the World Problem from the Standpoint of the New Economics*, p. 115. Elkin Mathews and Marrot.
41. Programme and membership application form in *New Britain*, October 1932.
42. Strathclyde University Archive T-Ged 9/1841.
43. Muir, W. (1968). *Belonging: A Memoir*, p. 168.The Hogarth Press.
44. NEG publicity leaflet at the New Atlantis Archive at Ditchling.
45. Quoted in Rigby, A. (1984). *Initiation and Initiative: An Exploration of the Life and Ideas of Dimitrije Mitrinovic*, pp. 109–10. Columbia University Press, New York.
46. Maxwell, D. W. (1932). *The Principal Cause of Unemployment*. Williams and Norgate; G. D. H. Cole (ed.) (1933). *What Everybody Wants To Know About Money: A planned outline of monetary problems*. Victor Gollancz.
47. Rigby, A. (1984). *Initiation and Initiative: An Exploration of the Life and Ideas of Dimitrije Mitrinovic*, p. 111.Columbia University Press, New York.
48. Advertising slogan in *New Britain*, 27 Dec 1933, p. 163.
49. Letter of Frederick Soddy to Winifred Gordon Fraser, 17 April 1934, New Atlantis Archive.
50. Rigby, A. (1984). *Initiation and Initiative: An Exploration of the Life and Ideas of Dimitrije Mitrinovic*, p. 1. Columbia University Press, New York.
51. Interview between the author and Ralph Twentyman. Forest Row, July 1991.
52. Interview between the author and Harry Rutherford. Ditchling, May 1991.
53. *New Britain* weekly, 9 May 1934, p. 748.
54. Interview between the author and Grace Rutherford. Ditchling, May 1991.

55. Letter of Frederick Soddy to Winifred Gordon Fraser, 3 October 1933, New Atlantis Archive.
56. For Odon Por, see Lectures under the auspices of New Britain, January 1934; articles, *New Britain*, 21 February 1934, 18 April 1934; for Fuller, see, for example *New Britain*, 16 May 1934, although his connections grew stronger in the 1940s.
57. *New Britain*, 1 (2), 52–3.
58. Mairet, P. (1981). *Autobiographical and Other Papers*, (ed. C. H. Sisson), p. 86. Carcanet, Manchester.
59. Selver, P. (1959). *Orage and the New Age Circle*, p. 57. George Allen and Unwin Ltd.
60. Muir, W. (1968). *Belonging: A Memoir*, p. 41. The Hogarth Press.
61. Watts, A. (1973). *In My Own Way: An Autobiography 1915–1965*, p. 109. Jonathan Cape.
62. Watts, A. (1973). *In My Own Way: An Autobiography 1915–1965*, p. 111. Jonathan Cape.
63. Watts, A. (1973). *In My Own Way: An Autobiography 1915–1965*, p. 109. Jonathan Cape.
64. Watts, A. (1973). *In My Own Way: An Autobiography 1915–1965*, p. 123. Jonathan Cape.
65. Watts, A. (1973). *In My Own Way: An Autobiography 1915–1965*, p. 124. Jonathan Cape.
66. *New Britain*, 15 November 1933, p. 816.
67. Letter of Winifred Gordon Fraser to Frederick Soddy, 12 April 1934.
68. *New Britain*, 18 April 1934, p. 688; letter of Frederick Soddy to Winifred Gordon Fraser, 12 April 1934.
69. *Principles and Aims of the New Atlantis Foundation*, p. 12.

CHAPTER 8
The darkest days

The period of Soddy's life from 1935 until after the Second World War was cryptically summed up in his obituary in *The Times*, 'Then after 1936, there was silence . . .' While not strictly accurate, this points to a problem in any attempt at a complete biography of Soddy. After 1936 he almost disappeared from the various areas so far discussed—science, monetary theories and New Britain—and, with some minor exceptions, the silence lasted until the late 1940s. Finally, there is almost no evidence for his activities during the whole of the Second World War. Unlike the other chapters of his life, the explanation for this is quite straightforward. The decade from 1935 was the worst of Soddy's life. It was the period during which his professional life ended, his hopes for peace and prosperity were shattered and his worst fears about the implications of his atomic discoveries were realised. Perhaps worst of all, however, his personal happiness was destroyed when his wife, Winifred, died. She became ill in 1935 and then died on 17 August 1936. This marked the end of a remarkably contented twenty years of marriage. As his friend Sir John Russell wrote in 1956, 'The death of his wife . . . was a blow from which he never fully recovered'.[1]

The effects on Soddy and his work were immediately disastrous. He turned from all his previous interests to a new and apparently absorbing study of mathematical formulae but although he seemed to be turning in a new direction, enough of his earlier interests remained that he took an original line throughout. He represented some of his findings in verse and attempted to produce mechanical models of others. However, before this, the first of his workings are to be found in the series of 'Mathematics Notebooks' which he kept from 1935 until about 1937 and then from 1941 to 1946. Page after page of these notebooks is covered with figures. The writing is unlike his usual style, becoming more heavily indented, with blacker and blacker ink.[2] Soddy's purpose behind these figures was the calculation of the solutions to cubic equations and he also developed a machine which he believed would mechanically solve these problems—possibly using his interest in joints for the workings of this. He had registered two patents for improving epicyclical gearing in 1931[3] and then he had attended an Exhibition of Hooke's Inventions in Oxford in 1935 and made notes on the 'Double Hooke's Joint'.[4] Together, these resulted in his patent for the 'Improvement in Hooke's Joints and

Couplings-Provisional' in 1938.[5] The second set of these notebooks is very like the first, full of interminable columns of figures in thick black ink.

He also set his mind to problems in geometry like the 'Kissing Circles' and the 'Hexlet' which he then put into poetry and published in *Nature*,[6] starting a minor correspondence of similar writings.[7] In 1941 he wrote a review entitled 'Qui s'accuse s'aquitte', of Hardy's *A Mathematician's Apology*, and then in 1942 and 1943 he published further mathematical works on the infinite harmonic series.[8] When he attended the meeting of Nobel Laureates in Lindau in 1955 he spoke on 'The Cubic Equation and a Machine which solves it'.[9] Then while sorting his papers in the 1950s for the biography written by Muriel Howorth, he put the notebooks aside for particular notice, and as we have seen, and left money for the development of this machine in his will.[10]

Although it is dangerous to stray into psychological theories for which there is little evidence, it does seem that Soddy's behaviour, at least so far as the calculations in the notebooks is concerned, can be read as a mechanical and absorbing way of avoiding the need to think about a distressing subject after the death of his wife, especially since Ernest Rutherford, one of those to whom he had been closest for over twenty years before their later estrangement, died unexpectedly in October 1937. While clearly less affecting than the death of his wife, this must also have increased Soddy's depression and removed a potential support, so that turning to the unemotional world of figures may have been a way of escape. This interpretation is strengthened by the announcement, in September 1940, of the death of his close friend since their school-days, Harold Cort Carpenter. Again, this coincides with a period of intense calculations in his notebooks. Together, these deaths marked the end of his connections with his youth. His wife, his parents, Beilby, Rutherford and now Carpenter had all died and he had no comparable relationships to replace them as support in his black moods.[11]

However, Soddy also felt that his mathematical work had some intrinsic importance and there is some support for this, not only in the pages of *Nature* during Soddy's life, but also in his correspondence with Professor Coxeter in the 1950s and more recently in the work on 'Computer Aided Research into the Geometry of the Triangle' undertaken by Adrian Oldknow.[12] Oldknow has not only worked on what he has named the 'Soddy line' but has rendered his findings into verse which reflects Soddy's original version. The use of a computer in Oldknow's investigations prompts curiosity as to whether Soddy's calculations might have borne fruit if he himself had had access to such technology. As in so many areas, it seems that Soddy's thoughts were anticipating later developments.

Soddy's retreat from personal contacts at this time was not just a psychological one. He also moved physically. He resigned his post at Oxford in 1936, at the early age of fifty-nine, and there is little sign that he was missed. In his correspondence there is a letter from A. W. Stewart, his successor at Glasgow, regretting his resignation and mentioning that Soddy's impression on students at Glasgow was still remembered, but there is nothing from his Oxford colleagues. Stewart's remarks about the Oxford dons that 'they have not been altogether congenial colleagues' well described the whole period from Soddy's arrival to his departure.[13] By February of 1937 Soddy seemed to be feeling that his resignation was perhaps rather precipitate. He wrote to his successor Cyril Hinshelwood to ask if he could have space in the laboratories since 'when I resigned it was with the idea of making way for a younger man rather than giving up my experimental work'.[14] Accommodation was found for him, and he was seen to go in and out until the outbreak of war, but although he employed 'a boy' to clean it,[15] there was little or no evidence that he actually did much research while there.[16]

This revival of interest in the use of a laboratory seems to have been brought about by a renewal of one of his earlier interests which, like his interest in Hooke's joints, dated from the 1920s. Now it was monazite sands, which he had last mentioned in 1922.[17] The precise reason for his renewed interest is not clear, but in August he wrote to the High Commissioner of India and to his Mineral Advisor about the Travancore Monazite Sands, referring to the renewed trade in monazite which was 'due to the demand for radio-active mesothorium'.[18] His notebooks record that from June 1937 to June 1939 and then from August 1939 to May 1941 he was involved in various experiments with monazite. However, the most important personal consequence of this interest was his decision to go himself to investigate the situation in India. In November 1937, he embarked on a six month trip to India, Malaya and Ceylon, partly for 'business', and partly because, when he had completed his world tour in 1903 he had believed that Ceylon was the most beautiful place on earth and now wanted to return and see it again.[19]

He left Oxford on 10 November 1937 and travelled via London to Marseilles where he embarked on the S.S. Oxfordshire for Cochin in India. He spent almost a month in India, in which he visited Cochin (where he experienced a rickshaw) and travelled south to Triandorm, the capital of Travancore, by steamer. He was fascinated by the country, writing of bullock teams ploughing 'under water'. He saw the canal systems, the lagoons and 'lovely green tropical scenery, everything at its best now, coconuts and palmyra palms, and paddy fields of the most delicate green' and commented, 'In a way this place is one of the wonders of the world.'[20] He collected specimens of sand and then moved on to Kandy and Colombo in Ceylon. While in Ceylon he saw the sun rise from the Holy Mountain, or Adams Peak, after a 'slog of thousands and thousands of steps' to the top.[21] He then went to Malaya via the Straits, passing Singapore, Tampan and Malacca, and returning to Calcutta. He finally returned home on 8 May 1938, calculating that his trip had cost fifty-five shillings a day.[22]

This spell of apparent happiness and even scientific work was short-lived. On his return he left Oxford, moving out to Knapp, a small house in the village of Enstone, where he spent the war years. This is presumably partly the consequence of his retirement, but it was also to do with his widowhood. After this, he never again lived the kind of comfortable lifestyle he had shared with his wife. There seem to be no accounts of his life at Enstone. He apparently managed to remain there for some ten years without visitors, or at least without any who have recorded the fact. After this, despite his reasonably comfortable means, he chose to live in some discomfort in Brighton in a small, dark house with only a housekeeper for company, sleeping in a small metal folding bed which was a remnant of his childhood.[23]

On his return from his Eastern tour, Soddy began investigations into his samples of sand. He seems soon to have abandoned notions of the radioactive potential of his samples, but began to investigate the commercial properties of them for the manufacture of gas mantles, one of the many products hit by the war. However, in the end it was decided that the whole proceeding needed too large a capital outlay and, although he continued his interest in the sands until 1940, nothing ever came of it.[24]

This seems to have been Soddy's final experience of practical science, and it is not at all clear that the 'scientific' aspects of the monazite were ever much more than an excuse for a much needed break and an excuse to return to Ceylon. Afterwards he wrote the occasional article about science, including 'Social Relations of Science',[25] and an address to the Science Master's Association after he was made their President in 1943,[26] but until the end of the war he published very little.

At a much less active level than during the early 1930s, Soddy continued an interest in monetary reform. However, this was in rather different kinds of organisation than the NEG, although, as with so much in the inter-war London world, there were strong connections and shared members, meeting places and ideas. It was probably *via* this method that Frederick Soddy came into contact with the Economic Reform Club and Institute which, for a time, shared the premises of the NEG in Gower Street.[27] The leader of what became the Economic Reform Club and Institute was Edward Holloway, who had been brought to ideas of monetary reform through contact with Arthur Kitson, with Eimar O'Duffy, who wrote for *New Britain* on monetary reform and published a book on the subject in 1932, and then with Orage and the *New English*

Weekly. Holloway was active in various movements which sprang up in the thirties whose aim was 'to bring many of these [monetary reform] movements together'.[28]

The first manifestation of this ambition was the Petition Council. The aim of the council was quite straightforward, to collect signatures to a petition concerning the abolition of poverty, and the removal of the main causes of economic warfare between the nations. Addressed to the King, the petition asked for some kind of judicial enquiry into the causes of economic difficulty. The petition received wide support throughout Britain and 'the Dominions' and the churches. From the basis of the Society a Petition Club was formed and premises acquired in Grosvenor Place, near Victoria Station. Although thousands of signatures were collected, the Petition movement ran out of steam in the late thirties and the outbreak of war made it redundant. However, the Club continued under the name 'Economic Reform Club' and found premises in another nearby house which provided a snack bar and a couple of bedrooms for members. It later added 'and Institute' to its title and finally merged with the Economic Reform Council. At its height in the late thirties, the membership of the Club rose to about two thousand. It continued, until the exigencies of wartime made it impossible, to provide some London premises for members and to hold regular dinners with speeches.

Unlike the NEG and New Britain, where the development of the individual and then of the group had been an important part of the programme, in none of these organisations was there any demand for personal involvement or belief in any area outside monetary reform. This unemotional approach seems to have appealed to Soddy at the time and he became one of the Vice-Presidents of the Club in 1936, an office he held until his death.[29]

Soddy published very little on monetary reform during this period. Most of what he wrote merely reiterated the views expressed in his books of the mid-thirties. There were two articles in 1935 in *Garvin's Gazette*, 'Money as Nothing for Something' in March and 'The Gold Standard Snare' in July.[30] In 1936 he wrote 'Is Science a Failure', in which he seemed to feel the answer was yes, that money was now a more important issue than science.[31] His report on the Prosperity Campaign's Conference at Digswell Park in August 1936, 'Money and the Constitution', was published by the Economic Reform Club and so was his '"The Budget", a synopsis in one hundred verses of the author's "Reformed Scientific National Monetary System"' in 1938.[32] In 1943 he wrote a little-noticed booklet, *The Arch-enemy of Economic Freedom*, reiterating his plans for monetary reform and attacking banks and bankers, and especially McKenna, previously Chancellor of the Exchequer, in his article for the *Banker's Magazine*, 'What is Banking'.[33]

This booklet was a closely-argued attack, but provided nothing new and lacked the immediacy and attention to contemporary events which had marked Soddy's earlier pamphlets. This was his last publication on monetary reform, although he continued to lecture on the subject.

He gave a number of talks to the Economic Reform Club, including 'Credit, Usury, Capital, Christianity, and Chameleons' in 1937.[34] He spoke to the Rotary Clubs at Aylesbury and High Wycombe in April 1939 on 'Money the mischief maker' and then he spoke to his most influential audience of the period, to members of the Parliamentary Labour Party, on 'Finance and the War', on 9 November 1940.[35] Unfortunately, but revealingly, no reaction to this can be found.

The most obvious cause of his silence was the outbreak of war. Soddy's efforts since Armistice Day in 1918 had been towards the peaceful uses of science and the reform of society to make that possible. Now he had quite clearly failed. He needed to decide on his next course of action. One possibility would perhaps have been some kind of involvement in the Allied war effort or the development of the atomic bomb. He was, after all, a genuine pioneer in the area. It was suggested after Soddy's death by Dr L. E. C. Hughes that 'he refused to come under security and assist in atomic work during the recent war'.[36] This has a ring of truth, especially in view of his difficulties with authority during the Great War and with the Oxford bureaucracy in the intervening period. Further, an important element in all his involvement with politics had been a commitment to peace. However, additionally, it has to be stressed that his science was, by this time, very out of date. Since the time when he had played an active role in atomic science, which was really before 1914, it had developed out of all recognition. Badash has described how radiochemistry had been 'suicidally successful' so that, by the 1920s, 'chemical studies of the radio-elements effectively ended'.[37] After this, the development of quantum theory took the study of radioactivity into theoretical areas inimical to Soddy's materialistic, hands-on approach to experimental science while the necessity for team work in the 'big science' of the thirties and later required co-operative efforts, like those of Rutherford's team at Cambridge, which demanded talents exactly opposite to Soddy's.[38] As his later resignation letter to the NEG concludes, he was better working alone, not as part of a team.[39]

His attitude here shows how far he had travelled, intellectually, from earlier colleagues and friends. Rutherford and Harold Carpenter, his closest friends amongst his generation of scientists, had of course died some years earlier. The younger generation of left wing scientists who, as we have seen, shared many of his aspirations, in this case felt very differently and wholeheartedly supported the national war effort from the very onset of the Second World War.[40] Their beliefs and intentions were set out in a statement issued by the Association of Scientific Workers in 1939:[41]

> It is a war in which science will play an important part, and it is necessary that our scientific resources should be used in such a way that they can be of the greatest effect in supporting the whole defence machinery of this country.

H. G. Wells, with whom Soddy had had much in common since 1913, was concerned throughout the war to set out a formal Declaration of the Rights of Man to ensure that, after the war, clear notions of democracy and how they could be achieved were ready to hand. Gregory, the editor of *Nature*, and by this time President of the British Association, was also associated with this, along with many others of the older generation who would not be taking an active part in the war. They produced a document which was widely disseminated and translated, and this led to the 1941 British Association Conference in London being given over to the topic with representatives of USSR, China, the United States and nineteen other nations. This generated two publications, a British Association account of the conference in its *Transactions* and then a version in a Penguin Special in 1942.[42] However, although 'most of the major Nobel Prize winners' took part in a transatlantic radio round-table discussion, there is no sign of Soddy attending.[43] Without some evidence, it is impossible to give reasons for his absence, but perhaps he had grown so dispirited after the end of the New Europe Initiative that he simply had no energy left for another fight.

However, the end of the war was to cause a considerable change. On 16 July 1945, the first 'atomic bomb', was exploded in the deserts of New Mexico.[44] On 26 July, the Potsdam Declaration was released to the press. On 6 August, 'Little Boy', the first operational atomic bomb, was dropped on Hiroshima. On 8 August, 'Fat Man', the improved atomic bomb, was dropped on Nagasaki. On 10 August 1945, the Japanese Emperor Hirohito finally surrendered. At Hiroshima in August 1945 and over the next five years, it has been estimated that 200 000 persons, or 54% of the population, were killed. At Nagasaki in August 1945 and afterwards, the figure was 140 000 persons, again 54% of the population. These figures compare with a ten per cent death rate after the fire-bombing of Tokyo.

These events affected people differently. Wells, having fought for peace and a democratic Europe after the end of hostilities for the whole duration of the war, was now 'too profoundly tired and unhappy' to speak about them on the BBC.[45] Soddy reacted in a diametrically opposite way. He came out of his self-imposed exile and began to speak and to publish again. His worst fears were at last realised:[46]

> When the 9 p.m. news broadcast on August 6 announced that ... the first

atomic bomb had destroyed Hiroshima, the writer experienced that same curious inversion of the order of events in time as, once before, when he fell climbing alone in the Rockies. If under such circumstance, the subject realises anything at all, he realises he is safe before he realises he is in danger.

It seemed that, now the worst had occurred, he could stop worrying and try to help reconstruct the world and repeat his warnings about the power inherent in scientific discoveries which should therefore be kept in the hands of scientists. On 18 August he published an article in *Cavalcade* explaining 'what exactly has happened':[47]

... a team of American and British scientists have successfully solved the problem of the release of atomic energy on a practical scale; that atomic bombs have been made at a cost of £500,000,000 in American factories; that the first was dropped on Hiroshima and the second on Nagasaki, smudging them out and the greater part (almost certainly, as an unofficial guess, all but a few), of their populations of over 500,000 people, and, within the week an offer to surrender has been made by Japan.

He went on to predict that 'the new power science has so light-heartedly and irresponsibly put into the hands of men drunken with conflict . . . is only too likely to prove a boomerang'. Here he showed remarkable prescience. Recent study on the making of the atomic bomb, of which there has been an enormous amount, has emphasised the extent to which two different motives fitted exactly together to enable the making and the dropping of the bomb. For all concerned there was primarily the need to end the war, and for victory over the enemy. This was the desire which had prompted Fritz Haber to work on poison gas during the First World War, even after he had seen his nitrogen fixation process used for the manufacture of explosives—the need to end the war as soon as possible by any means at all. However, in addition there was the reason put forward by J. Robert Oppenheimer, the director of the Manhattan Project. Immediately after the war he said:[48]

... the reason that we did this job is because it was an organic necessity. If you are a scientist you cannot stop such a thing. If you are a scientist you believe that it is good to find out how the world works; what the realities are; that it is good to turn over to mankind at large the greatest possible power to control the world and to deal with it according to its lights and values.

This had led to the making of the bomb. However, dropping it was achieved by 'turning it over' not to 'mankind at large' but to 'the hands of men drunken with conflict', the military whose sole aim was to end the war and the politicians whose primary aim, by the time of the Potsdam conference, was to ensure that the United States of America came out of the war as the most powerful nation. This determination meant that attempts by a minority of scientists, of whom Albert Einstein and Leo Szilard were in the vanguard, to share the knowledge gained through the research programme were ignored or rejected.

It is worth remembering at this point that Szilard had been inspired, to a greater or lesser extent, to begin his search for the chain reaction, which led to his involvement in the development of the bomb, by reading H. G. Wells's *The World Set Free*, with its dedication to Soddy. Throughout the war, Szilard had remained one of the scientists most aware of the potential social and political implications of the results of the research into nuclear power. He had collaborated with Einstein on occasions from 1939 to 1945, the last in a final attempt to achieve some freedom to publish the result of the research as an alternative to dropping the bomb. After the war, at the conference on Atomic Energy Control at the University of Chicago in September 1945, Szilard set out his fears for the future:[49]

We are in an armaments race. If Russia starts making atomic bombs in two or three years—perhaps five or six years—then we have an armed peace, and it will be a durable peace. But we will not have permanent peace at lesser cost than world government. But this cannot come without changed loyalty of people.

Here, in a few sentences, he summarised all that New Europe had attempted during the thirties. Here he also emphatically, if unknowingly, supported Soddy's notion of a *boomerang*. Without political process, the atomic bomb would remain a recurrent threat, liable to come back at the researchers. This is now awfully familiar and needs no explanation. We all now live with the constant threat of nuclear power and so understand one action of the boomerang. Yet it is worth noting Soddy's fears from at least the 1920s, if not earlier. In those awful days of 1945 all he had seen and predicted came true. As D. G. Soden wrote to *The Times* after Soddy's death,[50]

[In 1922] speaking of the probability of 'splitting' the atom and liberating atomic energy he [Soddy] said: 'Every scientific discovery in this direction is but another nail in the coffin of mankind.' These prophetic and tragic words have been with me ever since. There seems a possibility that their truth is being recognised now.

By September, Soddy was beginning to realise this possibility. He had had time to consider further the implications of the bomb and to realise that the boomerang effect he had mentioned in August was of immediate consequence. Because of the secrecy imposed on the work of the atomic scientists after Pearl Harbour, he had no certain knowledge, but he guessed that the bombs had been dropped without any 'intermediate stage of small-scale experimentation under carefully controlled experimental conditions', so there was only mathematical proof that they would not cause untold destruction. However, 'so far, at least, fortune has favoured the brave', the world survived the exploding of the bomb and so future peaceful use of atomic power could be anticipated so long as one drawback is overlooked, 'the effect on the human being of the rays generated'. He mentions the precautions taken in the medical use of radium to avoid radiation burns and points to the 'evidence of this in the uncannily delayed action of the "slight" burns resulting from the Hiroshima bomb'. His conclusion was surprisingly optimistic. Because the dangers of the rays were so great, he expected that atomic power could only be used where necessary precautions against the rays was possible, which was, as he had said in 1909, in 'thawing a frozen pole or making a garden of Eden' not against human beings.[51]

Soddy's August reference to Prometheus who 'stole fire from heaven and was chained for ever after to a rock in punishment' was equally prophetic. In America, the McCarthy years are sufficiently well documented that the future of Oppenheimer and others is so well known a part of our culture that it does not need describing here. In Britain, although the immediate post-war years had pointed to a highly auspicious beginning with a Labour Government committed to establishing the welfare state, with 'a new and widely shared consensus about the importance of science in national life' beginning to emerge, and with a newly politicised generation of scientists who turned to the Left for the political views, this was all short-lived. The effects of the Cold War spread from America and resulted in a precipitate decline of the British Left, and, especially of left wing British scientists.[52] As ever, the freedom of scientists was curtailed by those who paid the bills.

Soddy had been predicting these changes since August 1945 when he had asked what he regarded as the essential question:[53] *'Control? Who controls whom? Those who have this power control the whole world.'* In an account that is one of the best and most powerful things he ever wrote, he argues that control of nuclear power will probably go to scientists, but that scientists will be controlled by 'an international what-not'. However, had those in control over the last century, during which science had provided such opportunities, managed to

. . . control war, unemployment, the debauchment of the rural and urban population, artificial poverty in the midst of plenty, or any single one of the grave social disorders that the displacement by science of draught-cattle, human and animal, by steam and electric power, the tractor and the motor, has brought in its train?

[The answer was] No! relative to the times they live in, they are imposters, doing as they are bid by 'experts', imposters whatever age they live in, expert in exacerbating, exploiting, and even creating these very social disorders instead of abolishing them.

He goes on to argue, as he had been arguing for some thirty years, for an alternative form of control, for some better advisors 'than the broken reeds that wrecked the previous century of scientific achievements'.

Finally, Soddy pleaded, again as he had for decades, for control of science by scientists. He gave a brief history of the discovery of atomic power, emphasising the internationality of the scientists involved. Then like Szilard and Einstein he asks 'Why exactly is it that the scientists cannot be trusted?' Soddy's 'melancholy conclusion' was 'that those responsible for government fear and hate science because, by its capacity to render men economically free, it requires better men than they are to govern at all.'[54]

These articles seem to have renewed Soddy's interest in atomic science, both on its own and as a partner to his ongoing concern with monetary reform. In the article in *Cavalcade* he referred dismissively to 'the international monetary authority agreed on at Bretton Woods' and the 'proposal to socialise the Bank of England'. By this time, his disillusionment with any kind of government or controlling authority was so complete that, even though the reforms he had been advocating for the past twenty years were at last to be enacted, they would be ineffective. 'The false alchemy of the middle ages preserved as mystic and cabalistic wisdom by masonic and secret societies and their like, is at bottom the sole stock-in-trade of the people behind the ostensible Government, who hold the destinies of the nations in the hollow of their hands.'[55]

Over the next two years he continued to formulate these ideas and talk about them. In July 1947, he gave an illustrated talk to the New Europe Group in the British Academy Lecture Room,[56] and then in October 1947, brought his ideas together in an address 'An Independent Scientist's Views on the Economic and Political Possibilities of Atomic Energy for the Future' given to 'The Constitutional Research Association'.[57] This talk seems to have been sparked by a series of broadcasts earlier in the year by a number of experts on the atomic bomb and are in part, at least, dependent on a series of articles he was preparing on the history of atomic energy which appeared firstly in *The Engineer* and then were published in book form as *The Story of Atomic Energy*.[58] From this work, he gave a brief summary of the history of the development of atomic power. He ended with the thought that since a delay of ten years would allow the development of nuclear power by both major powers, the United States and 'The Communists', it might be better if the next war occurred immediately. This would lead to world domination by the USA, a view shared at this point by Bertrand Russell. Soddy's political shift to the right is revealed here when he refers to the communists as 'the pseudo-religious fanatics' whose avowed aim and objective has been world domination.[59] He then voiced his belief that a considerable amount of research was still needed before atomic power could be harnessed for peaceful ends. Then, with a masterly change of direction, he described his own efforts to reform the monetary system in the inter-war period, linking this very loosely to the subject of his lecture by explaining that 'in the advanced stage of rapidly approaching dissolution civilisation has reached' a simple solution like his own would probably be insufficient, but that until the money system in re-inverted and restored to its original function as the national instrument for distributing all that can be produced at a constant price level, science will continue to bring not peace and prosperity but frustration, wars and destruction over the whole world.

However, he seemed to recognise that this kind of explanation was no longer sufficient. The last war had, by its very occurrence, demonstrated that his efforts to prevent it had been inadequate. On the other hand, its ending had more than proved his fears about atomic energy to be right. This seems to have started a search for some kind of philosophy which would help him to come to terms with the consequences of his early work. He makes the connection himself by remembering his experiences at the St Louis Exposition in 1904. He remembered that he had seen there[60]

a gigantic human figure in cast-iron, many times life-size, representing the god Vulcan. Alongside of it . . . was a diminutive exquisitely carved figure in white marble . . . of the Christ . . . Now . . . it typifies the whole world about to be subjugated by atomic energy, in its obvious moral of the domination of the spiritual by the material.

However, he does not, as might have been expected, return to Christianity, to 'spineless modern religions', nor can he quite abandon logic so far as to embrace monetary reform with his old fervour. No matter how hard he tries, it is clear that his system was too simple for the post-war world. In this lecture he can find only 'the gods of the ancient world [who] have come back and demand our obeisance on pain of extermination'.[61]

He gave some more details of his views in the conclusion to *The Story of Atomic Energy*, the book published in 1949 which was his scientific swan-song. Here he again blames Christianity for the 'spineless and one-eyed attitude towards the problem now facing all humanity' which leads to an attempt to 'invert the world of fact and commonsense, to affirm that black is white, and escape into the unreal world of its own creation'.[62] He contrasted this with pre-Christian civilisations. In ancient Egypt, 'science was the secret of the ruling hierarchy', in Greece before Plato and Aristotle it was 'an essential part of culture and philosophy'. Therefore he advocated a return to an understanding of truth as understood by the Greeks. This was but a temporary haven—the philosophy of the ancient Greeks and Romans was too far from the concerns of the twentieth century to support him through the next few years.

The end of the war also marked a revival in Soddy's interest in monetary affairs. Reginald Wrugh remembers that he had bought the entire stock of Soddy's pre-war pamphlets from Colletts just before the outbreak of war since nobody, least of all Soddy, was interested in them. In the immediate aftermath of the war, Soddy, his interest renewed, was delighted to have them again.[63] The immediate cause of Soddy's renewed interest was the financial and economic policies of the Labour government. This interest was first expressed in an article in the *Torch of Truth*, 'Can a Nation ever be saved by Budgets and Ever-Increasing Taxation?'[64] This takes issue with Norman Crump, financial editor of *The Sunday Times*, who had set out the problem of increasing income and saving without causing inflation. Soddy argued, as usual, that there was a confusion in Crump's thought between wealth and debt, and he predicted that the problems over the next few years would echo those of the immediate post-First-World-War years. The second effect of the budget was more personal, although Soddy had hoped to illustrate a wider point.

A 'Special Contribution' had been levied in the budget as a part of the government's attempts to deal with post-war financial difficulties. When this fell due on 1 January 1949, Soddy refused to pay it on the grounds that it was an unconstitutional demand and because he had been looking for a way of challenging the legality of the existing money system.[65] He decided to appeal and, as 'the successive demands for payment were ignored', a summons was issued for a hearing in the High Court on 28 March 1949. For this, Soddy produced an affidavit and then attended the hearing. In his words,[66]

The scene and manner of conducting the case had little of the decorum and dignity one would naturally expect of a Court of Law, most of all a Court of the Supreme Judicature of the land, but it did strongly suggest an illustration from one of Dickens' novels. A rabble of litigants and their lawyers contesting tax demands were herded against a barrier behind which sat the Master.

Within two minutes, Soddy's case was dismissed as 'frivolous and vexatious' and leave to appeal was refused.[67] When he continued to refuse to pay, Soddy was declared bankrupt and finally, in March 1950, an order was issued against him for the debt and £13 costs.

Soddy had no doubt hoped to inspire others to follow his lead but, as Sir Richard Gregory, the past editor of *Nature*, said in a letter of support, no one else had either objected to the payment or questioned its validity. The reason, Gregory thought, was 'because most people in these days are invertebrate in uprightness of thought . . .' (spinelessness seems to be a feature of this period of Soddy's life).[68] As a result no support for Soddy was forthcoming from the general public. However, he was supported by his friends in the Economic Reform Club and Institute and by NEG who attempted to start a 'Soddy Society for Honest Money' and who published an account of his case. From 1948, when he had moved to Brighton, Soddy had been in fairly close contact with the Economic Reform Club again, especially with the secretary, Edward Holloway, but could do little at first since he had been unwell.[69] However, in 1951 he put his case to them in an address, 'The constitutional justification for resisting tax-payments', in which he explained 'every detail in the decline of our world-position can be simply explained by the corrupt money system which began here with the granting of the Charter of the Bank of England in 1694'.[70] He pursued a similar theme in 1950 when he spoke to one of the organisations set up almost as younger siblings of the Economic Reform Club, 'The Birmingham Paint, Varnish and Lacquer Club', on the subject of monetary reform.[71] Here he explained the urgency of his mission, that another war threatened between the 'two irreconcilable camps of communism and capitalism' which, because armed with atomic and biological weapons, threatened to exterminate civilisation altogether. However, instead of looking back to his 1945–7 lectures on the effects of science on warfare he now argued, as he had before the war, that 'the impending collapse' was 'the consequence of its false money system'.[72]

There is something almost pathetic in these speeches, because he refused to accommodate them to what had happened after the war. The nationalisation of the Bank of England was dismissed with a sentence, the development of nuclear power was ignored and all the terrors of the Cold War could, he suggested, be prevented by monetary reform. Even if his solution worked, a population suffering from the rigours of 'Utility' and rationing, homeless or making do, terrified by the prospect of an even more dreadful war, beginning to recognise the financial costs of the war just over, needed to hear proposals softened by sympathy and some understanding, not just 'I told you so'.

However, he never really followed the insights of the immediate post-war years, but instead turned towards the non-party-political conservative point of view. This can be seen particularly in his relationship with the NEG. As well as with the Economic Reform Club, after the war and especially after his move to Brighton, he renewed his connections with the New Europe Group. After the war, the rather few remaining members of the Group came back together and decided that new political or economic initiatives were pointless, but that the third area, the cultural, might be worth pursuing. To this end they formed a lecture society, the Renaissance Club, also known as the anti-Barbarous Renaissance Club of the New Atlantis. There were also the New Boadicea Club for women and the New Caractacus Club for men. The Club was described like this by one of its members:[73]

The purpose of the Renaissance Club is, as its title implies, to make more widespread the realisation that in the present crisis of human life nothing less than Rebirth is adequate. We are faced with such an unprecedented situation—science having brought to us the choice between almost unbounded wealth and leisure, if we had the courage to accept it, or racial suicide if we cannot change the whole basis of our life.

The Club existed for twenty years and organised more than 200 lectures on a variety of topics by a number of speakers, many of whom had no other contact with the NEG. However, there were also a number of the 'old guard' from the pre-war days, including Soddy, Major General J. F. C. Fuller, J. T. Murphy and the Duke of Bedford (previously the Marquis of Tavistock). On 1 July 1948, Soddy was one of the speakers at a meeting at the Swedenborg Hall for a general forum discussion on 'Initiative for British Renaissance', then on 6 August he spoke at another meeting at the same place with the same organisers to commemorate Hiroshima. Soddy's speech on this occasion showed a shift in his idea of himself. He began: 'I must make it clear that I am a pure scientist, working in the more inhuman side of science. On the side of social problems I have less claim to speak than any of you.'[74] This is a direct reversal of his pre-war position, and indeed the position held by him since the early 1900s. A possible reason for his shift is revealed in the next paragraph in which he said, 'The bomb was dropped contrary to the opinion of American Scientists had they been consulted'. More recent knowledge shows this to be largely wishful thinking, but it seems likely that Soddy's thinking about the history of research into atomic power and the subsequent responsibility for its use had returned to the formula he had previously rejected, that science is above politics, the scientist is merely a researcher, good or evil may come from any discovery but that is the responsibility of society. This interpretation is strengthened by his conclusion. After rehearsing the history of atomic power he described the building of the first atomic pile at Chicago University as 'an easily controlled, self-supporting chain reaction' which meant that 'man has the means of energy which make him into a God'. However, 'unless he lives like a God he will destroy himself'. Finally, Soddy suggested that

instead of commemorating Hiroshima we should celebrate the date when the experiment first proved successful ... Let us celebrate man's triumph over the problem, and not its first misuse by politicians, and military authorities.

It is a speech of which Rutherford would have been proud. It abdicates responsibility for the use of research, it keeps the scientist pure. The younger Soddy would have scorned such a position, but now he was old and tired. He was over seventy, as he reiterated in his speeches, he was the last survivor of his generation of scientists. He alone had lived to see the fruits of their work. The other survivor, although not a scientist, was H. G. Wells, but their friendship had faded away by this time and there is no record of post-war contact.

In February 1950, he lectured to the New Boadicea Club on 'Why a larger pay-packet now buys less than it did', possibly his most dispassionate lecture of the period.[75] In this he rehearsed his well known views on the necessity to reform the banks and money system and then discussed the role of the scientist. Here, somewhat curiously from one who had so articulately supported the social responsibility of science in the thirties, he suggested 'whether the use society makes of our brains is constructive or destructive lies less in our hands than in those of any other branch of the learned world. We are treated as butlers . . .' He then, and it should be remembered that this is the same Soddy as the one who had, forty years before, supported the education of women and their fight for the vote, finished with a condescending suggestion that 'women, with their age-old experience in the necessary art of debunking the posturing male, might be able to start it [the exposure of the fallacies within the international economic system] now with a demure feminine titter'.[76]

He was also involved with public initiatives of the NEG. In 1948 there was a 'memorable trip to Rome for the Congress of the European Union of Federalists'.[77] Two incidents from this trip have been described by his friend David Shillan to 'show something of the man in his later years'. The first was when he despaired of getting his point of view understood during a debate on a single European currency and so he 'rose in majesty and led a walk-out by the British delegation'. The other was 'a more intimate one'. After a visit to the Catacombs he commented, 'My Brother should have seen this', not because he was an archaeologist but 'because he is a Christian'.

Soddy had issued his book *The Story of Atomic Energy* under the publishing body of the NEG, Nova Atlantis. Despite the objections of the printers to his chosen format, the book was issued. Then there were difficulties with distribution which had been undertaken by Nova Atlantis. By December 1952, sales were so small that there was a possibility of the stock being taken over by a larger publisher. However, nothing was done and by January 1954, it was still the case that 'those ordering through booksellers have not been able to obtain it'.[78]

The days of the NEG were in any case numbered. What was thought to be the final meeting of the NEG, at which Mitrinovic made his last public appearance, became known as 'The Simpson's Lunch'. It was a 'Luncheon Press Conference' at Simpson's Restaurant in the Strand on 17 February 1950. The speakers were Frederick Soddy, Major-General J. F. C. Fuller and Dimitrije Mitrinovic, and they were to speak and answer questions on 'U.S.A., U.S.S.R., Great Britain and Europe: Their

Roles in the Present Situation'.[79] However, there was at least one encore. The distinction between them and the Renaissance Club was never clear when they were organising lectures. For example, a series of lectures held at the College of Preceptors in Bloomsbury in the Autumn of 1950 was under the auspices of the NEG whose address was Renaissance House.

After this, the Renaissance Club continued to organise lectures, changing its name to The New Atlantis Foundation after the death of Mitrinovic in 1953. Mitrinovic had moved to Richmond, Surrey in 1948 and the rest of the Group had followed when the lease of Norfolk Lodge, near the top of Richmond Hill, became empty. Mitrinovic's health had worsened after an attack of pneumonia in 1947 and after the move to Richmond he was less and less inclined to go out and was eventually confined to bed. He spent his time ensuring that his companions understood as nearly perfectly as possible the essence of his ideas. He died on 28 August 1953.

Just as the public face of NEG, and Mitrinovic's health, suffered from the war and its aftermath, so did their financial situation. During the war, Mitrinovic had been forced to sell some of his paintings and other works of art,[80] although there were enough left to decorate the 'red drawing room' at Richmond Lodge which so impressed one of the younger members with its 'Epsteins and other masterpieces'.[81] Richmond Lodge had only been bought with the help of a friendly bank manager and the total assets of all the members of the Group, although it proved a particularly sensible investment in view of the inflation of house prices in that part of London. The proceeds from its eventual sale helped the (by then) ageing members of the Group to continue their work after they moved to Ditchling in 1978.[82]

Soddy continued his association with the New Renaissance Club, exchanging books, staying at the Charlotte Street premises when in London, and being involved with the preparations for the lecture series. In 1950 proposals were made for the Group to renew their interest in monetary reform through the foundation of a Soddy Society, and although nothing came of it, they did run three lectures on the subject. However, Soddy also had another role. The finances of the Group had always been precarious and by the early fifties things had worsened. It is unclear whether this was caused by a general decline because of dwindling support, of whether the move to Richmond had proved the final blow, but by January 1952, the Group needed to borrow £2000. They approached Soddy who described his position in the most depressing terms. 'The tax-case has left my interest in public affairs near zero-point and I have been thinking of withdrawing from the fight altogether and taking a rest from the money issue.' However, although he said that 'I personally do not see any prospect of the Group, as its finances are at present run, paying any loan back', and so was unwilling to lend any money, he offered to take over the mortgage on the Charlotte Street house. He did this, at a cost of £680 5s 0d, and then, in March 1952, became a Director of 'the Nova Atlantis Property Company Limited', formed 'to bring the management of our various properties on to a sounder basis'. Despite these efforts, financial pressures were so great that 96 Charlotte Street had to be sold in August 1952.[83]

In his August letter, Soddy again stressed how disillusioned he was and gave this as his reason for limiting the assistance he was prepared to offer to the NEG. He felt that

the whole political situation is so hopeless not only here but throughout the world that any further effort to reform it is a waste of time and energy and one can only hope that there will be nothing worse.

But everything at the moment points to things getting worse before they can get better and for my part I feel like resigning from any further reform movements, including the New Europe Group, as simply a waste of time. I feel too that by nature I am and work better on my own rather than as a member of a group.

After this, it is not surprising that he wrote on 10 October 1953 commiserating on the death of Mitrinovic, refusing any further financial help and resigning his position as President of the New Europe Group. He 'wishes to express my appreciation of and thanks for the kindness and friendliness of all the members of the Group. However my own interest in social topics had evaporated with increasing years.' His resignation was accepted on New Year's Day 1954, and marked the end of twenty years of association with the Group.[84]

Notes and references

1. Anon. (1957) 'In Commemoration of Professor Frederick Soddy', p. 3. New Europe Group, Ditchling.
2. Professor Frederick Soddy Papers, Bodleian Library, Oxford, MS Eng Misc b174, b183 and b184.
3. Copies in the Royal Society *Personal Records*, Soddy F. 11.
4. Professor Frederick Soddy Papers, Bodleian Library, Oxford, MS Eng Misc b180.
5. Copy in the Royal Society *Personal Records*, F. 11.
6. Soddy, F. (1936). The Kiss Precise. *Nature*, 20 June 1936, p. 1021; The Hexlet. *Nature*, 5 December 1936, p. 958.
7. Various authors (1937). The Bowl of Integers and the Hexlet. *Nature*, 9 January 1937, p. 77; 23 January 1937, p. 154; 6 February 1937, p. 251; see also the account of this in Gardner, M. (1979). *Mathematical Circus*, pp. 32–5. Penguin, Harmondsworth.
8. Soddy, F. (1942). The Summation of Infinite Harmonic Series. *Proceedings of*

the Royal Society, A **179**, 377; (1943). The Three Infinite Harmonic Series and their Sums. *Proceedings of the Royal Society*, A **182**, 113.
9. Soddy, F. (1956). *The Cubic Equation and the Machine which Solves it*. Speech given in 1955. New World Publication.
10. Soddy, F. Note attached to 'Mathematical Notebooks' in Professor Frederick Soddy Papers, Bodleian Library, Oxford, MS Eng Misc b174.
11. The mathematical notebooks are in Professor Frederick Soddy Papers, Bodleian Library, Oxford, MS Eng Misc b174; for Carpenter's death see Professor Frederick Soddy Papers, Bodleian Library, Oxford, MS Eng Misc b120, b181 and *The Times*, 6 September 1940, p. 6, and *Tribute*, 26 September 1940, p. 7.
12. Oldknow, A. (1994). The Soddy Line of a Triangle: Harmonies on a Line of the Kiss Precise. Unpublished MS. Chichester. See also his Computer Aided Research into the Geometry of the Triangle. *Math Gazz.*, November 1995. I am also grateful to Dr Oldknow for the information about Professor H. S. M. Coxeter.
13. Professor Frederick Soddy Papers, Bodleian Library, Oxford, MS Eng Misc b189.
14. Frederick Soddy to Hinshelwood, 6 February 1937. Professor Frederick Soddy Papers, Bodleian Library, Oxford, MS Eng Misc b189.
15. Professor Frederick Soddy Papers, Bodleian Library, Oxford, MS Eng Misc b188.
16. Hinshelwood to Muriel Howorth, 1957. Professor Frederick Soddy Papers, Bodleian Library, Oxford, MS Eng Misc b189, 236.
17. See Chapter 6 above.
18. Professor Frederick Soddy Papers, Bodleian Library, Oxford, MS Eng Misc b171.
19. Accounts of this tour are in Professor Frederick Soddy Papers, Bodleian Library, Oxford, MS Eng Misc b170 19, b171 39, 406. See also *Pioneer*, pp. 238–9.
20. Frederick Soddy in a letter to Hilda Beilby, 5 December 1937. Quoted in *Pioneer*, p. 307.
21. Frederick Soddy in a letter to Hilda Beilby, 27 December 1937. Quoted in *Pioneer*, p. 308.
22. Own Cash Book EASTERN TOUR 10 xi 37–8 iv 38. Professor Frederick Soddy Papers, Bodleian Library, Oxford, MS Eng Misc b171.
23. Interview between the author and Margaret Lewis. Summer 1991.
24. Letters from Frederick Soddy about he monazite sands between 1937 and 1940 are contained in Professor Frederick Soddy Papers, Bodleian Library, Oxford, MS Eng Misc b171.
25. *Nature*, 30 April 1938, pp. 184–5.
26. Soddy, F. (1943). Science for Rulers. *The School Science Review*, November 1943, **95**.
27. What follows is taken from Holloway, E. (1986). *Money Matters: A Modern Pilgrim's Economic Progress*, especially Chapters 3 and 10. The Sherwood Press.
28. Holloway, E. (1986). *Money Matters: A Modern Pilgrim's Economic Progress*, pp. 21–4. The Sherwood Press.
29. Holloway, E. (1986). *Money Matters: A Modern Pilgrim's Economic Progress*, p. 94. The Sherwood Press.
30. Soddy, F. (1935). *Garvin's Gazette*, March and July 1935.

31. Soddy, F. (1936). Is Science a Failure? *Radiography*, July 1936, **11** (19).
32. There are reprints of these in the Royal Society *Personal Records*, Soddy F. 11.
33. Soddy, F. (1943). *The Arch-Enemy of Economic Freedom: What Banking is. What first it was and again should be.* Published by the author at Knapp, Enstone, Oxon.
34. *Pioneer*, p. 309.
35. Wise, L. *Frederick Soddy, Great Money Reformers No. 3*, pp. 9–10. (n. d.) Holborn Publishing and Distributing Company.
36. Obituary in *The Times*, 28 September 1955.
37. See the article by Badash in Kauffman, G. B. (ed.) (1986). *Frederick Soddy (1877–1956): Early Pioneer in Radiochemistry*, pp. 38–9. Reidel, Dordrecht.
38. Rhodes, R. (1986). *The Making of the Atomic Bomb*, Chapter 6. Penguin Books, Harmondsworth.
39. Letters between Frederick Soddy and the NEG, 10 October 1953, 30 January 1954. New Atlantis Archive.
40. Werskey, G. (1978). *The Visible College*, Chapter 8. Allen Lane.
41. The A.S.W. and the War. Undated but October 1939. Quoted in Werskey, G. (1978). *The Visible College*, p. 262. Allen Lane.
42. Smith, D. C. (1986). *H.G. Wells: Desperately Mortal*, Chapter 17. Yale University Press, Newhaven.
43. Smith, D. C. (1986). *H.G. Wells: Desperately Mortal*, p. 437. Yale University Press, Newhaven.
44. The information in this paragraph is taken from Rhodes, R. (1986). *The Making of the Atomic Bomb*, pp. 733–5, 630 and *passim*. Penguin Books, Harmondsworth.
45. Smith, D. C. (1986). *H.G. Wells: Desperately Mortal*, p. 467. Yale University Press, Newhaven.
46. Soddy, F. (1945).The Atomic Age. *The Scottish Field*, October 1945.
47. Soddy, F. (1945). The Moving Finger Writes . . . *Cavalcade*, 18 August 1945, pp. 8–9.
48. Quoted in Easlea, B. (1983). *Fathering the Unthinkable: Masculinity, Scientists and the Nuclear Arms Race*, p. 90. Pluto Press.
49. Quoted in Rhodes, R. (1986). *The Making of the Atomic Bomb*, p. 753. Penguin Books, Harmondsworth.
50. *The Times*, 25 September 1955.
51. These ideas were published in the October article in *The Scottish Field* cited above, which had been written in September.
52. For a discussion of these matters see Werskey, G. (1978). *The Visible College*, Chapter 8. Allen Lane.
53. Soddy, F. (1945). The Moving Finger Writes . . . *Cavalcade*, 18 August 1945, pp. 8–9.
54. Soddy, F. (1945). The Moving Finger Writes . . . *Cavalcade*, 18 August 1945, pp. 8–9.
55. Soddy, F. (1945). The Moving Finger Writes . . . *Cavalcade*, 18 August 1945, pp. 8–9.
56. Soddy, F. (1949). The Story of Atomic Energy and its Lessons. Lecture to the New Europe Group, 15 July 1949. Professor Frederick Soddy Papers, Bodleian Library, Oxford, MS Eng Misc b180 f.123.

57. Soddy, F. (1947). *An Independent Scientist's Views on the Economic and Political Possibilites of Atomic Energy for the Future*. Address to The Constitutional Research Association, 30 October 1947. The Constitutional Research Association, n. d.
58. Soddy, F. Articles on the history of atomic energy in *Engineering*, 3 October 1947–14 May 1948; Typescript, *Modern Alchemy*, or *Real Alchemy and the Release of Tomic Energy* (1947). Professor Frederick Soddy Papers, Bodleian Library, Oxford, MS Eng Misc b185. This is a draft for the articles cited and for *The Story of Atomic Energy* (1949). The Nova Atlantis Publishing Company Limited.
59. Soddy, F. (1949). *The Story of Atomic Energy*, p. 11. The Nova Atlantis Publishing Company Limited.
60. Soddy, F. (1949). *The Story of Atomic Energy*, p. 14. The Nova Atlantis Publishing Company Limited.
61. Soddy, F. (1949). *The Story of Atomic Energy*, p. 15. The Nova Atlantis Publishing Company Limited.
62. Soddy, F. (1949). *The Story of Atomic Energy*, p. 128. The Nova Atlantis Publishing Company Limited.
63. Interview between Reginald Wrugh and Stuart Laing. 1989.
64. Soddy, F. (1948). Can a Nation Ever Be Saved By Budgets and Ever-Increasing Taxation? *Torch of Truth*, May 1948.
65. This account is taken, except where otherwise noted, from Soddy's own version, which he entitled 'Frederick Soddy Calling All Taxpayers'. It was published in a mimeographic form by The Soddy Society for Honest Money, 96, Charlotte Street (the offices of the New Europe Group).
66. 'Frederick Soddy Calling All Taxpayers'. Published in a mimeographic form by The Soddy Society for Honest Money, 96, Charlotte Street.
67. Further details from David Shillan's speech to the Soddy Trustees, *Seventh Report of the Frederick Soddy Trust*, November 1983, p. 7.
68. Professor Frederick Soddy Papers, Bodleian Library, Oxford, MS Eng Misc b189, 255.
69. See letters from Holloway to Soddy in Professor Frederick Soddy Papers, Bodleian Library, Oxford, MS Eng Misc b189.
70. *Pioneer*, p. 241.
71. Soddy, F. (1950). Monetary Reform as a Preliminary to All Reform. Address to The Birmingham Paint, Varnish and Lacquer Club, 12 January 1950.
72. Soddy, F. (1950). Monetary Reform as a Preliminary to All Reform. Address to The Birmingham Paint, Varnish and Lacquer Club, 12 January 1950, p. 3.
73. Information about the New Europe Group and letter quoted in Rigby, A. (1984). *Initiation and Initiative: An Exploration of the Life and Ideas of Dimitrije Mitrinovic*, p. 172. Columbia University Press, New York.
74. The text of this speech is in the New Atlantis Archive.
75. Soddy, F. *Why a larger pay-packet now buys less than it did*. Lecture to 'The New Boadicea Club', one of the women's initiatives of the New Europe Group, February 1950. New Boadicea Club/NEG, n. d.
76. Soddy, F. *Why a larger pay-packet now buys less than it did*. Lecture to 'The New Boadicea Club', one of the women's initiatives of the New Europe Group, February 1950, p. 20. New Boadicea Club/NEG, n. d.
77. Details of this trip from Shillan, D. (1980). 'Geography benefits from a chemist'.

Geographical Magazine, November 1980, pp. 99–101; personal communication from Harry Rutherford at Ditchling.
78. Letters of 6 December 1952 and 30 January 1954, Frederick Soddy to Mrs Wrugh at Nova Atlantis Publishing Co. Ltd., 96, Charlotte Street, London. This was, at the time, the home of the offices of the New Renaissance Club.
79. This information is from the invitation they issued.
80. Rigby, A. (1984). *Initiation and Initiative: An Exploration of the Life and Ideas of Dimitrije Mitrinovic*, p. 171. Columbia University Press, New York.
81. Interview between the author and Grace Rutherford. Ditchling, Summer 1991.
82. Interview between the author and Harry Rutherford. Ditchling, 1990.
83. Letters between Frederick Soddy and the New Europe Group, January–August 1952. New Atlantis Archive.
84. Letters of Frederick Soddy to the New Europe Group, 10 October 1953, 30 January 1954. New Atlantis Archive.

CHAPTER 9

The post-war world

The end of the Second World War, as we have seen, left intact little of what Soddy had fought for in the inter-war period. The virtual disintegration of the New Europe Group and the manifest failure of monetary reform to influence the post-war Labour Government seems to mark the beginning of a period of defeat. However, two different areas of interest and concern changed this. Firstly, and vitally, he rediscovered a positive commitment to science, and to atomic science in particular. Secondly, he returned to his love of travel. The rediscovery of these areas was sparked by two remarkable women, both noted for their determination in following through their plans, but one far more successful than the other. Other than occupying aspects of Soddy's life at the same time, there is no connection between these two, and so although they both need to be considered for the effect they each had on him, they will be dealt with separately. When Soddy met her in the late forties, Margaret Tatton had been running some part of the Le Play Society from the mid-twenties, and while she had shown her efficiency in organising their tours, her methods of doing so had not always been popular. One member of the group wrote, 'Beware the Tat whose gentle voice/ All will obey—they have no choice/ The velvet glove will quite conceal/ The clutching hand with grip of steel.'[1] Muriel Howorth was just as determined to achieve her ends, but less successful in fulfilling them. She had developed an interest in atomic science and attempted to create an organisation within which non-scientists could learn about its potential, but this, for reasons discussed below, was never very successful, and was short-lived. She involved Soddy in this and also decided to write his biography.

Her reasons for this are never given in the remaining letters and accounts of her life at this time and it seems at first glance a somewhat extraordinary thing to undertake, since he had not been involved in any of the recent developments in the area. However, after the revelation of the power of atomic energy through the bombs, the public was fascinated with the subject and desperate for any information. In England, this was difficult to come by. Unlike the US authorities, the British government had decided on a complete embargo on any news of developments of a peaceful or military kind, so any news came through the American press where information was freely given. In part, this was because there was, at the end of the war, nothing to report in England. The scientists actively involved in work on atomic or nuclear energy had moved to the American projects during the war, and England had expected merely to share in those developments afterwards. In contrast, the Americans, who felt responsibility for the bomb, were now emphasising possible peaceful uses of this massive energy, and the possible safe harnessing of what had been unleashed. Various factors like the developing Cold War continued to bring out this national difference, even after the beginning of British research initiatives, so that it was not until 1948 that Harwell was allowed to admit the press.[2]

In addition to this deliberate policy of silence, there was another difficulty inherent in any British publicity, which was the lack of British nuclear scientists. As we have already seen, the earlier generation were by now nearly all dead; most of the younger ones were either involved in the production of wartime non-atomic military necessities, actually fighting, or for some 200 of them, involved in the American project. After the war, any one involved in any project connected with atomic energy was placed under the jurisdiction of the Official Secrets Act.

Thus, anyone attempting to publicise the subject in Britain clearly faced enormous difficulties. For Howorth, the discovery that Soddy was still living, of sound mind and in reasonably good health, must have seemed a wonderful stroke of good fortune. Here was someone who had been actively involved in the early days of discoveries, who had taken part in the 'greatest discovery ever made by man in the history of Time, that of the NATURAL TRANSMUTATION OF THE RADIOACTIVE ELEMENTS',[3] who was English and had never been connected with the American project and who was one of the few English Nobel Prize winners. The question of nationality was clearly important to her. In her edition of his *Memoirs*, she refers to 'this now-matured English chemist'.[4] Also, she soon found that she could present him as a martyr, the scientist who had been forgotten by history. In addition, as some of those who met her have suggested, she was to become 'genuinely fascinated by this famous (infamous) personage in her midst', for she found he was living in Brighton. However, this correspondent continues, 'I would be surprised if her interest in Soddy would have been so strong were it not for the aggrandizement that might accrue to her.'[5]

Aspects of Muriel Howorth which might have explained her initial involvement in this project are and must remain an enigma, since she left no papers and her career both before and after Soddy's death has proved impossible to trace. There is no evidence of any earlier writing or similar expertise. By her own account, during the war she was employed first at the Ministry of Information, then at the Royal Aircraft Establishment at Farnborough. While employed here, either in the last months of war or the first of peace, she received a letter 'asking if I would like to interest myself in atomic energy, of which the Ministry of Supply had set up a department'.[6] Although she claims she had no previous knowledge of the subject, the RAE had long-term interests and research in the development of atomic weapons.[7] However, she immediately began reading in the library of Shell Mex House where the headquarters of the department was situated. In the course of this reading 'one book I singled out called *The Interpretation of Radium*' especially because the author 'had the welfare of the world at heart', a novel idea to Howorth who had always believed that the scientist had little concern for the application of his discoveries. Howorth's chronology here is confused. The Ministry of Supply had taken over the responsibility for atomic power in October 1945 and so she must have been a government employee at that point. However this seems to have been short-lived since by 1948 she had founded, with her husband, 'Major Humphrey Howorth MA Cantab', the 'independent' 'Institute of Atomic Information for the Layman' which claimed Albert Einstein as its first president. It is possible, although not admitting of any certifiable proof, that this institute was set up with government funding. Certainly, Howorth and her husband had connections with some part of the American atomic energy establishment through the journal *Atoms for Peace Digest*, produced by the United States Information Service and the American Embassy in London, in which Howorth tried to persuade Soddy to publish his articles.[8] Through this publication, Mrs and Major Howorth had also got involved with importing irradiated seeds from America which they hoped to persuade lay-persons in England to grow and to chart their results in the hope of finding useful mutations.[9]

By the time that he was contacted by Howorth in the early 1950s, as we saw in the previous chapter, the bombing of Hiroshima and Nagasaki had begun to force Soddy back into contact with the world around him. He had started to lecture and to write on the implications of scientific energy, and he had reissued his book *The Story of Atomic Energy*. He would no doubt have been delighted to find a new forum for his activities, especially since this was just after his tax case, while he was still suffering from the depressing effects of its failure and moving towards resigning from the NEG. Despite this, he was not, initially, in favour of Howorth's proposals.

The first record of any contact between them is a letter which she wrote to him at the end of 1952, sending him materials about her 'Institute of Atomic Information and related bodies'.[10] However, although interested, his reaction was far from favourable. While 'entirely with you in approving the dissemination of scientific knowledge to the laymen, I am poles apart from you in disapproving entirely as to the manner in which this is being done.' The science exhibitions at South Kensington and the Dome of Discovery at the Festival of Britain 'ought to have been at the funfair at Battersea Park' and, from the literature she had sent,

the efforts with which you are associated are of the same category. They tend to make science a laughing stock with out any worthwhile result, save to give an excuse to a number of fashionable folk to feel they are in the van of progress.

As might have been anticipated from this, he refused the offer of the presidency of her organisation, neither would he go to their dinner.[11]

This reaction would have dismayed many, but Howorth was a stubborn woman, as their correspondence increasingly reveals. She apparently wrote back accusing him of rudeness, for which he later apologised and, perhaps to demonstrate his remorse, he agreed to attend one of her meetings at the end of February, to meet those involved and to talk for half an hour about the personal side of his work with Rutherford.[12] After this agreement, Howorth met Soddy in the lounge of the Grand Hotel in Eastbourne and he spent the whole of lunch-time telling her his history of atomic energy, by now a very well rehearsed tale. She seems to have been genuinely fascinated by the story but also took the opportunity afforded by a relatively mellowed temper and 'suggested, over coffee, that we should, together, write the story of his scientific investigations'.[13] Howorth's powers of persuasion must have been displayed to the full here because, despite his earlier relegation of her ideas to the funfair, he now agreed to her suggestion.

By 27 January there were plans for the production of a volume of his memoirs and within two days he was writing about the various technical details and offering to travel to Eastbourne, where she lived, for another meeting.[14] They met and corresponded frequently over next few years until his death. The results of the meetings were the production of two books, the first volume of his memoirs, *Atomic Transmutation: The Greatest Discovery Ever Made: Memoirs of Frederick Soddy, F.R.S., Nobel Laureate* which was published in December 1953 and *Pioneer Research on the Atom: Rutherford and Soddy in a glorious chapter of science: The Life Story of Frederick Soddy M.A., LL.D., F.R.S., Nobel Laureate*, issued in 1958 after his death. There were also booklets, including *Science and Philosophy: Extracts from the Works of Frederick Soddy*, containing fifty-two 'Sunday Morning Reflections', the reading of which 'can project a day of calm reflection and unflagging incentive to scientific achievement'.[15] As well as these publications, Soddy attended meetings of the Institute, gave lectures and very soon became its president. In 1953 he gave some lectures: the first, 'Just Fifty Years Ago', on the 28 February 1953, the 'Jubilee' of the original discovery, with Rutherford, of transmutation, told the history of the discovery and was printed as an appendix to the *Memoirs*. There was then a series of 'three popular experimental lectures' in October, November and December at University College, London, on 'The First Quarter-Century of Radioactivity', again to commemorate the jubilee of the original discoveries.[16]

In letters to Howorth, Soddy complained that these events were, almost entirely, ignored. While he was used to this in regard to his monetary theories—as he said, the '*Press* . . . in the circles in which I am accustomed to move we nickname the Suppress'—his scientific works had almost always received favourable attention and this had continued after the war. Fritz Paneth in *Nature*[17] had said of his 1949 book, *The Story of Atomic Energy*, 'Prof. Soddy, long retired from active research and after years of scientific silence, has devoted himself to presenting a complete history of the main discoveries.' In this he had succeeded, especially for the general or student reader, but there was more to the book than that. Paneth put the work into a specific context:

To-day there is widespread discussion of the responsibility towards the community of men of science and particularly experts in radioactivity; but a perusal of Prof. Soddy's non-chemical writings of no less than thirty years ago shows how strongly he felt the duty to fight for a better order of things . . . This spirit forcefully pervades the whole book.

Now the situation seemed to be different. He wrote to Howorth of the 'complete absence of any review of your book about me' and, 'as I expected, yr. letter to the D.T. did not appear, as you say in it you mentioned my name—quite enough to ensure its suppression by our Suppress'.[18] He developed an elaborate theory to explain this suppression, but one obvious explanation seems to have been ignored. This was that none of the works or lectures was particularly deserving of review, especially of a good one. This explanation is quite compatible with various facts. For example, the *Memoirs* had received at least one review as it was noticed in *Nature*, in June 1954, but the notice was far from favourable. The anonymous reviewer explained that the book was in part based on the lecture which Soddy had given to the Institute of Atomic Energy for the Layman in February 1953 in which[19]

Prof. Soddy describes in no modest vein the series of events and experiments in which he participated, during the early part of the century. As a kind of introduction to the speaker, Miss Howorth has written these memoirs of Prof. Soddy in which she manages, in the course of what can best be described as a newspaper interview by an over-admiring reporter, to recount mostly in Prof. Soddy's own words his early life and his experimental work up to 1904.

Rather more unfortunate is the memory of a then student at University College, London who attended one of the winter lectures on 'atomic energy'. He reported, 'Soddy's talk was woefully off the mark for his audience. We expected politico-sociological stuff about the peaceful atom, and we got a rambling disorganised lecture on the isotopes, stable and unstable, and his role in the Rutherford/Soddy collaboration . . .'[20]

However, Soddy had come to believe that there was a 'conspiracy of silence' against him, and that this was the result of deliberate actions by some person or persons from Cambridge, unnamed but referred to as the Cambridge or Trinity 'claque'. He said, in what seems excessively aggressive language,[21]

By now, with the complete absence of any review of your book about me, you may be realising a little more the power of suppression—absolutely the only weapon the Cambridge school are now left with—and which has to be fought by trying to make it impossible for them to continue it, by exposing what is in my view nothing less than an organised conspiracy to falsify the history of science in their favour.

The reason for this attitude is obscure. As recently as 1950, in the review mentioned above, Paneth had begun, 'The name Frederick Soddy must be familiar to all men of science as the founder, with Ernest Rutherford, of the theory of radioactive disintegration and as the discoverer of isotopy.'[22] This certainly does not show any sign of Soddy being regarded as anything other than a respected, if now elderly, member of the scientific establishment. However, despite this, Soddy's role as an 'atomic pioneer' certainly was (and perhaps is) underestimated. While the existence of a conspiracy is unlikely, there are signs that Soddy's contribution to science was increasingly overshadowed by Rutherford. Norman Feather, for example, felt it necessary to write in his forward to Trenn's account of the Rutherford/Soddy partnership that their collaboration, far from being dominated by Rutherford as conventional accounts suggested, was one of 'equals'. This is of course borne out by Professor Trenn's research.[23] Whatever the rights and wrongs, Soddy's worries continued to grow and by September 1953, he was certain that some kind of attack was mounted on him and that it was of recent origin. He wrote, 'it is, of course, only since the War and the atomic bomb that a deliberate attempt has been made by the Trinity College Cambridge claque to leave out all reference to me and ascribe it all to Rutherford.'[24] By February 1956, he was referring to 'the ring that takes every opportunity to falsify the scientific history either in the interests of Jewry or of Trinity College Cambridge'.[25]

It seems possible that, deliberately or not, Howorth was responsible for Soddy's fears. The Howorths had originally moved to Eastbourne because Major Howorth, 'a life member of Trinity', had five college friends living there, and clearly some of them kept in touch with their Alma Mater.[26] From remarks in a letter written by Soddy to Howorth, it appears that some of the Howorths' Cambridge friends had been talking about Soddy and the gossip had been passed on.[27] Without knowing the nature of this story, its importance cannot be estimated, but it seems the only sensible explanation for his growing feelings of antagonism towards Cambridge. These feelings then began to spread to other areas of his life. He was asked to deliver a lecture to the Students' Society at Glasgow University and, when the physics and chemistry professors turned up, instead of being flattered by the attention, he read it as a threat which he overcame by not allowing it to cramp his style. This was at the university where his popularity had been highest, where he had made the discovery for which he had been awarded the Nobel Prize and where his personal life had been so happy. However, the paranoid feelings grew until, possibly under Howorth's influence, he developed a full-scale theory about the suppression of his part in any scientific discovery. Worst of all, the long-dead Rutherford, who had written a perfectly friendly letter to Soddy just before his death in 1937, was blamed for a large part of it. By 14 February 1956, Soddy was not content with attacking the 'Cambridge claque', but had added Einstein, who he all but accused of plagiarism, and talked of an 'anti-Einstein group to expose his plagiarisms'. He then added 'Jewry' to his list of enemies.

Beyond these attacks on fellow scientists, Soddy began to strike at his closer friends. In 1953 he decided that his book, *The Story of Atomic Energy*, should be passed from the New Europe Group to Howorth for distribution as he 'could not be bothered with Nova Atlantis' and this further separation from his old friends is shown by his writing to Howorth that he 'does not know what Shillan [a member of NEG] may or may not have been saying about you'.[28] Even Howorth herself was not safe. He criticised her efforts on his behalf, saying, 'Your own efforts to publicise my work are quite unnecessary and mistaken, just because you are in the lay fraternity and not an "elite" science bug [sic]'[29] and, later, 'As a scientific man believe me it does me no good to be in any way "lionised" by you, and a 2nd part of your "Memoirs" would be intolerably dull for the G.P.[general public]'[30]

In view of the way the collaboration with Howorth seemed to affect his temper, it is perhaps as well that the projected second volume of memoirs was abandoned. However, although remembering and being reminded of all his grievances gave rise to his reputation as 'a recluse . . . impatient of society . . . steak-faced white haired rebel . . . awe-inspiring'[31] and to his nickname amongst his friends as Professor Porcupine, Soddy also gained some very real pleasures from his association with Howorth.

He also enjoyed his attendance at the reunions of Nobel Laureates. As we have seen, his contacts with scientists had been minimal through his Oxford period and during the war, so to meet them again was for him an interesting and delightful experience. He regularly attended the series of conferences at Lindau in Germany for Nobel Laureates which began in 1951. Soddy attended the 1952 conference, when Irene Joliot-Curie also attended,[32] and 'it was a very happy meeting and I was asked by Prof. Mattauch to repeat my Lindau lecture on Isotopes on the way home at his Max Planck Institute at Mainz.'[33] As well as giving the lecture, he was able to collect some money, 65 DM, from the institute which overcame the restrictions on currency taken out of the country and provided a comfortable trip home.[34] In 1953 and 1954 he was accompanied by Muriel Howorth who enjoyed the proceedings enormously, especially since they provided an opportunity for her to satisfy her desire for self-aggrandisement. She was later to stress 'over and again the importance of their *joint* attendance at the Bodensee meetings'.[35] Soddy lectured at both these meetings, in 1953 on 'Isotopes', and in 1954 on 'Wider aspects of the Disintegration Theory'.[36] Soddy attended for the last time in 1955 with Mrs Beilby, his sister-in-law, when Howorth was too ill to go with him. On this occasion he took his mathematical machine for solving cubic equations to show to his colleagues. This was the year of the Mainau Statement. This statement was signed by eighteen Nobel Prize winners from Europe, the USA, and Japan, including Soddy, Otto Hahn and George Hevesy, and contains their appeal against the use of nuclear weapons. They said:[37]

We do not deny that today the peace of the world may be maintained by the fear of these deadly weapons. Yet we feel that it would be self-deception if governments should believe that over a long period the fear of these weapons will prevent war from occurring. Fear and tension have too frequently produced war. Likewise, it would be self-deception to believe that minor conflicts could always be settled by the use of traditional weapons. In extreme need, no warring nation will deny itself the use of any weapon that scientific techniques can supply.

All nations must bring themselves to the decision by which they voluntarily renounce force as the last recourse in foreign policy. They will cease to exist if they are not prepared to do this.

As this statement suggests, Soddy's views had returned to his earlier fears about the effects of atomic energy which he had been voicing since before the First World War. Similarly, Soddy's other new interest of this period pointed back to his long-term interest in travel. He had had a passion for hills and mountains sparked by his trip to the Lake District in his undergraduate days which, as we saw, was reinforced by his time in Canada. This had been shared by his wife to the extent that they are said to have spent an exciting winter's afternoon climbing Snowdon, miscalculating the time needed and so finding themselves on the summit in the dark, in a snow storm, trying to make tea on the stove they had carried. They also had regular, if less spectacular, holidays in the Alps, often with Winifred's family, and less frequent trips to Scandinavia.[38] As well as this, as we have seen, Soddy could relish the experience of travel in climates hotter than this, being fascinated by the local inhabitants and ways of life different from those to which he was accustomed. His return to Ceylon was an apt demonstration of how he could use travel as a method of escape.

The war had prevented any further excursions abroad, but during the early 1940s, Soddy made contact with the Le Play Society, an organisation which provided much needed companionship both in England and for travel. In the process it also firmly contradicts Soddy's statement that he preferred to be alone, since all accounts agree about his enjoyment of sharing his experiences. The Le Play Society of the post-war years was a descendent of the Sociological Society founded by Patrick Geddes with the help of Victor Branford in London in 1902.[39] Supporters in these very early days included H. G. Wells, Sidney Webb and Hobhouse. The influence behind this was Le Play, who had been a French mining engineer of the nineteenth century who had carried out what would now probably be described as sociological studies of the communities to which his work inspecting mines took him. His best known work, *Les Ouvriers Européens*, was published in 1855 with a second, more developed version in 1879, in which he demonstrated his method of investigating a society through the three fundamental determinants, *Lieu, Travail, Famille* (Place, Work, Family). Although this work was not published in England and so had little effect on this side on the Channel, Geddes came across it while in Paris and developed the theory into his formulation, Place, Work, Folk, thus widening the units of study. The result of the Le Play influence was an emphasis on field trips, or as Branford said, 'Let's go and see for ourselves'.[40] Geddes turned this into academic study 'aimed at making sociology an experimental science' through 'sustained questioning with its interplay of induction and deduction, observation, working hypotheses, renewed observation and experiment'.[41] The International Visits Association was developed before the First World War to arrange visits which would make this possible.

After the war, in 1920, the Sociological Society moved into new headquarters in London, Le Play House in Belgrave Road, which had been purchased by Mr and Mrs Victor Branford and let to the Society at £50 a year.[42] Here, the Society grew both in numbers and importance under the leadership of F. J. Adkins, Alexander Farquharson and Margaret Tatton. It had a dual purpose at this time, proving a suitable environment for study, for lectures and discussion, and also to organise study tours. This is where Lewis Mumford had met Frederick Soddy in 1920, who was possibly attracted by the proposal to form a study circle of the Sociological Society on 'Social Finance: an enquiry into methods by which credit may be diverted to social ends'.[43] For the next ten years the Society existed successfully, drawing up schemes of survey, organising the trip and in some cases printing the results in the *Sociological Review*. By this means it both influenced and reflected the growth of geography as an academic discipline and fulfilled a need for students and teachers to travel and to take part in fieldwork. This side of its work attracted increasing numbers of students and, to cater for their needs, the Student Group of Le Play House was established.

By 1930 the two kinds of activity in the Society were proving uneasy bedfellows. The Sociological Society itself, under Alexander Farquharson, became the Institute of Sociology, and it, together with the *Sociological Review*, remained very much under his control, moving out at the outbreak of war to Malvern and then to Ledbury and finally, after his death, becoming closely associated with the University of Keele.[44] Meanwhile, the tours were becoming more and more successful, and needed more and more organisation. From 1924 Margaret Tatton had been responsible for them and from 1927–8 this section of Le Play House had a semi-autonomous existence as the 'Foreign Work Committee of Le Play House', later shortened to Le Play Tours.[45] Tatton was paid around £200 a year as organiser by 1929, in which year she had accounts in progress for ten tours going to Corsica, the Tyrol, Spain, Bohemia and Geneva, amongst other places. By 1930 the list was more exotic and included North Africa, Finland and the Dolomites.[46] At this point the Society split and the Le Play Society, as it was henceforth known, moved to new premises in Gordon Square. The Student Group remained within the Le Play Society, and this helped to strengthen links with the courses run by Miss Charlotte Simson in the Cotswolds in field work and local survey.

Thus, throughout the 1930s, study tours were organised by or under the auspices of the society at a number of levels. There were, at the simplest, those in the Cotswolds. Then there were the student tours, designed for cheapness and accessibility to the numbers of geography students and young teachers. These were remembered with affection and stories of four in a bed, 'roughing it in the Carpathians, or sleeping on the floor of a school-room in an Albanian village' or less romantically, camping in Scotland, give a sense of the attraction for the younger members.[47] It is less easy to describe the other tours because there are no longer those around who remember them. However, throughout this period there was an emphasis on the *study* element of the tours. They could be of two kinds. Groups might settle in an area for a close survey: 'the physical setting, the ethnography, the settlement, the high farms, the village, its life and sociological structure, the school and the Church; statistical details of the population were collected, and a vegetation survey was made.'[48] These reports were edited and then published by the society. The second type of visit was typified by that which Sir John Russell led to the USSR in 1934 when they gained a general overview of the country by starting from Leningrad in the north, crossing the country in a south easterly direction to the Caucasus, then crossing the Ukraine and returning to Leningrad, all in the company of an Intourist guide.[49] Great pains were taken on all these trips to visit parts of the country usually inaccessible or unknown to ordinary travellers, local experts were found to explain local ways of life, traditions and the geography and, while accommodation was more comfortable than that offered to the student groups, it was often, by nature of the area visited, extremely basic and always as inexpensive as possible. For example, a trip to Russia in 1926 lasted thirty days and cost from forty five guineas, depending on the kind of accommodation.[50]

The other activity of the Society during the 1930s was the annual conference which, from 1932–8, was held in London in the Christmas vacation. Reports of tours and student surveys were made at these meetings, but they also included outside speakers who lectured or led discussions on questions of current interest, but excluded party politics.[51]

In a more limited way, these activities continued during the war. Trips were still made, but now to the highlands of Scotland. The annual conference moved out of London to Oxford. The headquarters of the Society could not continue to be in London so their various possessions were dispersed, to a school in Braughling, to Rothampsted and to the University of Nottingham, but with the remainder housed by Margaret Tatton and her sister in their house at Birling Gap in Sussex.[52]

Greater changes occurred after the war. The annual conferences at Oxford continued and efforts were made to continue the tours. However, restrictions on travel and inflation combined to limit visits to North-West Europe, and to prevent the detailed surveys which had been an earlier feature of the Society's work. By this time another, more fundamental difference was materialising. While, in the early days, large numbers of students had joined the Society, the need for this part of their work had declined. University courses now included fieldwork; of more direct relevance, the differences between the Students' Group of the Society and the main body were making themselves felt. By September 1946 'the committee had accepted the necessity for the separation of the Students' Group from the Le Play Society'. By December of that year the Geographical Field Group had emerged as a successor to the Student Group.[53] The result of these changes was that[54]

... membership of the Le Play Society was not reinforced by an influx of youth, was getting older, and more prone to demand the first-class accommodation and comfortable travel which made prices prohibitive to younger folk. The advertisement of the last foreign trip, to Dalmatia in April 1959, is significant. It had centres at Belgrade and Dubrovnik and involved 'studies of towns, peasant settlements, scientific and industrial institutions and archaeology; an interesting experience is looked for, and we hope it will be a wonderful holiday.'

This was the Society with which Soddy came into contact in the during and after the war. Just how he came into contact with Le Play is unclear, but there were some links between them and New Britain which might have made the introduction for Soddy. They had had a common root and president in Patrick Geddes and so it is likely that they had common members, although only one can be so identified: David Shillan, a close friend of Soddy in latter years, was a member of both organisations. Margaret Tatton thought that there was sufficient similarity for her to have advertised the Tours in *New Britain* in 1934. In addition, as this work has attempted to demonstrate throughout, the world of 'intellectual', politically active reformers in London between the wars was a small one.

However, the meeting took place, and it led to what was probably Soddy's most contented and happy relationship with any organisation. Unfortunately for the purposes of this work, 'Happy are the people whose annals are boring to read.' In this there were none of the excitements, the arguments, court cases, discoveries and publications of his earlier life, or even of his association with Muriel Howorth. From those who accompanied him, we know that he went to Sweden in 1949, and to the Tyrol in 1950, where 'an Austrian university woman who was an experienced climber ... managed to get the Professor up to very considerable heights',[55] In contrast, at Easter of the following year he was a member of Le Play Society East Sussex Meeting based in Eastbourne. He enjoyed the scenery and some of the history, but terrified the woman who was allocated the seat next to him on the coach by talking about what she remembers as 'bi-metallism', about which she knew, and chooses to know, nothing.[56]

Two other trips exemplify Soddy's relationship with the Le Play Society. There was a trip to Germany, at the conclusion of which he invited a group of members and their partners to be his guests on a trip on a steamer down the Rhine. This generous action became typical. Where appropriate, he would invite a group of friends to extend their visit at his expense, to see other nearby places or just to relax. One of these attempts, however, nearly failed. By 1955, he had developed a desire to see the Northern Lights and invited various people to accompany him on a Norwegian steamer which would travel along the Norwegian coast to the Northern

Cape for New Year's Eve. There, it was hoped, they would be able to celebrate the New Year and watch a display of the Lights. However, all except one of his invitations was refused on the grounds that it would be too cold and dark. However, a party of four was eventually assembled and a most successful and enjoyable trip was the result.[57]

The person who accepted the original invitation was Margaret Lewis, whom he had met through the Le Play Society and to whom much of the happiness of these later years must be attributed. She was a geography teacher, and this is how, as a student between the wars, she had come across the Society. She taught, by the 1940s, at a girls' school in Bexhill, and escaped from there to walk on the South Downs with Soddy, and to be taken out for meals when starvation seemed imminent. They also had at least one holiday apart from the Le Play when they walked in the Alps. This trip demonstrated that, while Soddy was now cheerful, he remained obstinate. On their way up to one of the mountain refuges, Soddy insisted that they stop for tea, despite the lateness and the distance remaining. By the time this was made on the stove he always carried on such expeditions they had to rush, and Soddy had some kind of seizure. Luckily, help was available and he was taken to the refuge. Then on the way down he demonstrated his complete lack of any sense of direction by trying to go the wrong way. Unlike most of his acquaintances in his past, Margaret Lewis seemed able to deal with such episodes with equanimity, and this seems to be the basis of their friendship. The rhymes she composed for him sum up her feelings,

> To talk
> Of Argon
> Jargon
> To some disciple on a walk
>
> To be
> Unique
> A freak
> Of Nature, who cannot agree
>
> With you—
> Or me
> To be
> Persuaded that anything is true
>
> Except
> The works
> And quirks
> Of one F.S.—the sole adept.

Soddy also attended the Le Play New Year Conferences throughout this period where it was 'not uncommon to find artists, scientists, historians, classical scholars and women of affairs ... some most interesting exchanges of views are possible. The breadth of interest appealed very much to Soddy.'[58] He enjoyed the social life at these conferences, and, from the armchair by the fireside which he made his own, perhaps imitating Mitrinovic who had done the same at the early New Britain lectures, also enjoyed the lectures and discussions.

Soddy found more than enjoyment, deep though this had been, in the Le Play Society, as is made clear in his will. He had suffered from increasing ill-health during the 1950s (which involved operations) and had clearly begun to consider the future. He had already been generous to friends and especially those who had worked for him in any capacity, giving money to laboratory staff. However, he now came to realise the extent to which the state, having already bested him in the matter of the Special Contribution, could follow him beyond the grave by claiming death duties. He therefore made individual gifts of £500 each, the maximum allowed without incurring duties, to a number of friends.

Frederick Soddy died on 22 September 1956 after a last operation at the beginning of that month. Before this operation, during the summer of 1956, Soddy called on Peter Bunker, then a young solicitor in Hove, to discuss the settlement of his affairs. Soddy needed an operation which was to take place within the next few weeks, and which he had decided would prove fatal. He was therefore anxious to organise his affairs before he kept his appointment at the Avenue Clinic, Hove. The first part of his arrangements was straightforward, consisting of gifts to family, friends and those with whom he had worked, which were to be presented as soon as possible so as to avoid death duties. He had made his will in which he left a substantial amount, £36 476 net, which he divided according to his perception of the needs of those associated with his various interests. Firstly, £1000 was left to Sir Richard Southwell FRS to develop and market his machine based on Hooke's joints, and his own version of the machine was left to the Museum of the History of Science at Oxford. Secondly, £1000 was left to Muriel Howorth, along with his papers and the copyright of his published works, 'in the hope, but creating no trust in the matter that she might be able to complete his memoirs'. The bulk of his estate was, however, to be used to set up an educational trust, to fund expeditions broadly along the lines of the Le Play Society. The workings of this Trust will be considered in the next chapter.

The range of these interests reflects the extent to which, even in his last years, Soddy remained an enigmatic and complicated figure. His tetchiness and his obsession with the conspiracy theory of the history of science contrasted with the benign figure in the memories of many who knew him. His death, marked by generous *Times* obituaries, saw him numbered among the great men of English science. Despite his own fears, most of the tributes fully recognised the importance of his work with Rutherford and Ramsay and all mention the work on isotopy and the Nobel Prize.

This view has been supported and, indeed, reinforced by more recent histories of science. The standard works on Rutherford and Trenn's work on the Soddy/Rutherford collaboration fully recognise the centrality of the young Soddy's contribution to the early stages of atomic science. Margaret Gowing's three-volume account of the development of atomic energy in Britain since 1939 also begins with a fulsome recognition of Soddy's role in the discovery of transmutation. In 1986, George B. Kauffman edited a collection of essays, based on the papers given at a Soddy Session at the Edinburgh History of Science Conference in 1979 and largely concerned with Soddy's scientific work. Further, the bibliography of this book shows a powerful awareness of Soddy's importance in the history of science.

The present work, insofar as it has added to this material, supports that view. Despite his difficulties and anger, Soddy was clearly an outstanding figure of early twentieth-century science. However, the period after the Great War still presents the academic establishment with many problems. Evan a basically sympathetic review in *Nature* of the Kauffamn collection describes his career as 'bizarre' and talks of him having 'almost no impact in his new sphere of activity'.[59] However, it has been a contention of this study that Soddy's feelings on science not only grew from his own experimental work but were far-sighted and, above all, consistent. This prescience has been recognised by Brian Easlea and more recently by Professor Mansel Davies, but the firmness with which Soddy clung to his dark vision has not really previously been appreciated. Further, we have already seen how a new generation of scientists, particularly in the environmental and social sciences, have come to see Soddy's theories of energy as of central importance.

All this suggests that the tunnel vision of scientific histories which measure achievement solely in terms of this or that laboratory or this or that grant leave many questions unanswered. There remains, though, Soddy's work on economics. Social Credit, the criticisms of bi-metallism and the demands for the reform of the baking system are now long forgotten parts of our intellectual history. Yet, as I have argued earlier, it is easy to underestimate the historical importance of a movement with the enormous clarity of hindsight. It is by no means clear that, if we could stand now in 1928 or 1935, it would be obvious whether Soddy and Kitson or Maynard Keynes held the key to the future. Further, New Britain and New Europe, for all the eccentricities of parts of their history, are much more in the mainstream of the intellectual history of the inter-war period than later events would suggest. Their theories of European Federation are now, forty years or more after they were published in *New Britain*, receiving serious consideration.

Nor was Soddy's interest in monetary theory as out of line with his interests in science as conventional accounts of his life suggest. His move to economics was prompted by a fundamentally humanitarian instinct which sought to avoid the terrifying Pandora's Box which splitting the atom seemed to be going to open. He brought his scientific method to bear on the questions of social organisation, and, like many of his generation, settled on money as the literal root of all evil. Only by tearing up this root and planting a fresher and better stock could science be preserved and protected for humankind. As he wrote in 1938,[60]

The effect of scientific discovery on the whole framework and content of sociology, economics, and politics has been with me a major preoccupation for nearly twenty years. Everywhere in the teeth of every sort of ignorance, I have witnessed the realisation of what the scientific civilisation might be, but for the procrustean bed of a private, upstart, and frustrating monetary system.

In the years after the war, as we have seen, both his worst nightmares and his greatest hopes suddenly seemed possible. The bombs dropped on Japan and then tested again and again in the last years of his life showed the true horror of what he had first glimpsed in Glasgow and Aberdeen before the Great War. Yet there lurked, behind all that, a greater and better vision. To Soddy, the generation of electricity by atomic power must have seemed like the first steps towards the Garden of Eden, which again he had seen in the 1900s. Yet he remained remarkably prescient. Just as we have come now to question the use of atomic power for peaceful purposes, Soddy, in 1955, wrote to Muriel Howorth:[61]

Isotopes of course are only a small part of the possible peace time uses of transmutation, the ultimate provision of almost unlimited qties. of energy in the ultimate future is much more important. Here, I merely object to the thing being prematurely induced—like the ground-nut scheme by ignorant politicians and vested monetary interests. Its too early to try and compete with the conventional sources—coal and oil. Also it is likely to be disastrous both to the human race and the climate the dirty way it is at present recklessly done, with no thought or care for the safety of the world and its inhabitants.

However, it is his last sentence in this letter which somehow encapsulates all that was best and most visionary in Soddy's science, 'a century hence will still probably be too early to attempt making the whole world a smiling Garden of Eden'.

In these last years, though, he found some kind of happiness and peace. In the company of Margaret Lewis and the Le Play Society, he seems to have gone back almost to a golden age of the world before 1914 and before the horrors of gas and bombs. In the expeditions in England and elsewhere we sense a very different Frederick Soddy, the rambler, the walker and the lover of the Lake District, who married Winifred Beilby in the last years of an old world. It is for this reason that he sought to turn the money from a Nobel Prize in Chemistry, which perhaps fathered the unthinkable, into the ways and means for a younger generation to learn, know and love the world.

Notes and references

1. From Margaret Lewis, copy in the author's possession.
2. This discussion relies on Gowing, M. (1964). *Britain and Atomic Energy 1939–1945*, pp. 50–5. Macmillan.
3. *Memoirs*, p. 6.
4. *Memoirs*, p. 7.
5. Letter from Professor T. J. Trenn to the author dated 29 December 1991. In the author's possession.
6. This information comes from her introduction to *Pioneer*.
7. See Gowing, M., assisted by Arnold, L. (1974). *Independence and Deterrence. Britain and Atomic Energy, 1945–52. Volume 2. Policy Execution.* Macmillan.
8. Copy of this in Professor Frederick Soddy Papers, Bodleian Library, Oxford, MS Eng Misc b173 f18; see also letters of Frederick Soddy to Muriel Howorth, 2 December 1953 and 12 December 1953.
9. Howorth, M. (1960). *Atomic Gardening for the Layman, passim.* New World Publications.
10. The letters to and from Muriel Howorth are from the photocopies which Trenn made of Soddy's letters while they were in his possession. They are now deposited in Professor Frederick Soddy Papers, Bodleian Library, Oxford, MS Eng Misc b171. However, as explained in the Introduction, the completeness of this series cannot be relied upon.
11. Letter of Frederick Soddy to Muriel Howorth, Christmas Day 1952.
12. Letter of Frederick Soddy to Muriel Howorth, 14 January 1953.
13. *Pioneer*, p. 16.
14. Letter of Frederick Soddy to Muriel Howorth, 27 January 1953 and 29 January 1953.
15. Howorth, M. Preface to *Science and Philosophy: Extracts from the Works of Frederick Soddy*. No publisher, no date, but contains a 1956 photograph of Soddy.
16. This description is from *Isotopy: Lectures by Soddy and Cranston*, p. 1. New World Publications, n. d.
17. Paneth, F. A. (1950). Review of Frederick Soddy's *The Story of Atomic Energy. Nature*, 11 November 1950, pp. 799–800.
18. Letters of Frederick Soddy to Muriel Howorth, 12 December 1953 and 22 February 1955.
19. Anonymous review of Muriel Howorth's *Memoirs. Nature*, 19 June 1954, p. 1159.
20. Letter from John T. Scott to the author dated 23 January 1991. In the author's possession.
21. Letter of Frederick Soddy to Muriel Howorth, 12 December 1953.

22. Paneth, F. A. (1950). Review of Frederick Soddy's *The Story of Atomic Energy*. *Nature*, 11 November 1950, pp. 799–800.
23. Trenn, T. J. (1977). *The Self-Splitting Atom: The History of the Rutherford-Soddy Collaboration*, p. 7. Taylor and Francis Ltd.
24. Letter of Frederick Soddy to Muriel Howorth, 8 September 1953. Professor Frederick Soddy Papers, Bodleian Library, Oxford, MS Eng Misc b171 287.
25. Letter of Frederick Soddy to Muriel Howorth, 14 February 1956.
26. *Pioneer*, p. 114.
27. Letter of Frederick Soddy to Muriel Howorth, 22 December 1953.
28. Letter of Frederick Soddy to Muriel Howorth, 22 December 1953.
29. Letter of Frederick Soddy to Muriel Howorth, 12 December 1953.
30. Letter of Frederick Soddy to Muriel Howorth, 7 September 1955. Professor Frederick Soddy Papers, Bodleian Library, Oxford, MS facs 129 f238.
31. Beaton, C. and Tynan, K. (1953). *Persona Grata*, p. 87. Wingate.
32. Professor Frederick Soddy Papers, Bodleian Library, Oxford, MS Eng Misc b171 27.
33. Professor Frederick Soddy Papers, Bodleian Library, Oxford, MS Facs 129 f224; letter of Frederick Soddy to Travers, 9 October 1952.
34. Professor Frederick Soddy Papers, Bodleian Library, Oxford, MS Eng Misc b189 f258.
35. Letter from Professor T. J. Trenn to the author dated 29 December 1991. In the author's possession.
36. *Pioneer*, p. 256.
37. Professor Frederick Soddy Papers, Bodleian Library, Oxford, MS Eng Misc b171 27.
38. Professor Frederick Soddy Papers, Bodleian Library, Oxford, MS Eng Misc b171 39, 406.
39. Except where otherwise stated, the information about the history of the Le Play Society comes from Beaver, S. H. (1962). The Le Play Society and Field Work. *Geography*, **47**, 225–40.
40. Quoted in Beaver, S. H. (1962). The Le Play Society and Field Work. *Geography*, **47**, 233.
41. Russell, Sir John (1960). The Le Play Society. The Presidential Address at the concluding conference of the Society, 9 to 13 April 1960, p. 5. Published as a report by the Trustees of the Frederick Soddy Trust.
42. University of Keele, Branford Papers, VB 205.
43. University of Keele, Branford Papers, VB 188.
44. University of Keele, Branford Papers, not numbered. Farquharson, A. (1954). A History of the Sociological Society. *Le Play House Bulletin*, **14**.
45. University of Keele, Branford Papers, VB 206.
46. University of Keele, Branford Papers, VB 162.
47. Uncatalogued material from 'Map Librarian' (apparently S. H. Beaver's papers). Draft typescript, University of Keele Archive.
48. Russell, Sir John (1960). The Le Play Society. The Presidential Address at the concluding conference of the Society, 9 to 13 April 1960, pp. 6–7. Published as a report by the Trustees of the Frederick Soddy Trust.
49. Russell, Sir John (1960). The Le Play Society. The Presidential Address at the concluding conference of the Society, 9 to 13 April 1960, p. 7. Published as a report by the Trustees of the Frederick Soddy Trust.

50. Russell, Sir John (1960). The Le Play Society. The Presidential Address at the concluding conference of the Society, 9 to 13 April 1960, p. 13. Published as a report by the Trustees of the Frederick Soddy Trust.
51. Russell, Sir John (1960). The Le Play Society. The Presidential Address at the concluding conference of the Society, 9 to 13 April 1960, p. 10. Published as a report by the Trustees of the Frederick Soddy Trust.
52. Russell, Sir John (1960). The Le Play Society. The Presidential Address at the concluding conference of the Society, 9 to 13 April 1960, pp. 10–11. Published as a report by the Trustees of the Frederick Soddy Trust.
53. Wheeler, P. T. (1967). The Development and Role of the Geographical Field Group. *The East Midland Geographer*, **4** (27), 186.
54. Uncatalogued material at the University of Keele.
55. Statement from Margaret Tatton, FRGS. In *In Commemoration of Professor Frederick Soddy M.A., LL.D., F.R.S., Nobel Laureate, formerly President of the New Europe Group who died 22nd September 1956*, p. 13. New Atlantis Group, n. d.
56. I am grateful for this information to Mr H. B. Harral, formerly Librarian of Eastbourne College, and to Mr and Mrs Underhill who took part in this trip. Mr Underhill formerly taught at Eastbourne College.
57. From interview between the author and Margaret Lewis, London, August 1991.
58. Russell, Sir John (1960). The Le Play Society. The Presidential Address at the concluding conference of the Society, 9 to 13 April 1960, pp. 12–13. Published as a report by the Trustees of the Frederick Soddy Trust.
59. *Nature*, 24 April 1986, p. 691.
60. *Nature*, 30 April 1938, p. 78.
61. Professor Frederick Soddy Papers, Bodleian Library, Oxford, MS Facs 129 ff 238–9.

Epilogue: The Frederick Soddy Trust

As we saw in the previous chapter, Frederick Soddy wished that the major part of his estate should be used to fund an educational trust which would continue the work of the Le Play Society. He had thought through the question of the constitution of the trust very carefully. He suggested that the fund was to be administered by a Board of Trustees consisting of four members, and awards could be made only on the basis of a unanimous decision. If such a decision proved impossible, then the matter was to go to arbitration. The youthful and courageous Peter Bunker pointed out that the appointment of an arbitrator would be impossible with four trustees and suggested that the Professors of Geography from Oxford, Cambridge and London Universities should be asked to appoint the arbitrator and their decision was to be binding. Bunker went on to indicate that this was a somewhat cumbersome procedure, and that a more satisfactory process would be an odd number of Trustees, to be bound by a majority decision. Despite Soddy's initial very fierce reaction to the suggestion, he agreed and accepted the pattern which eventually emerged of an odd number of Trustees. Then an even more formidable task was put upon the solicitor. Soddy asked that he, Peter Bunker, should be the independent chairman of the Trust because he would be able to keep Margaret Tatton in order, something Soddy had never been able to achieve because he found Tatton 'terrifying'. This seems extraordinary since, as this description from David Shillan shows, Soddy was not a person to be trifled with: 'You must imagine that when this massive figure, who kept some of the extraordinary good looks of his earlier days, rose up in wrath it could be rather daunting for anyone who had enraged him.'[1] Bunker thought Soddy 'a formidable, not to say frightening individual', so the prospect was uncomfortable to say the least. However, he agreed to the task and so was appointed together with four others. Margaret Lewis, David Shillan and Margaret Tatton had all been close to Soddy during his later years. The fourth Trustee was Soddy's niece, Marjorie Soddy.[2]

In 1957, after Soddy's death, the Trustees met for the first time and wrote to *The Times*, *The Manchester Guardian* and *The Times Educational Supplement* to announce the foundation of the Frederick Soddy Trust. In *The Times*, the announcement read:

The Trustees of the £25,000 fund established under the will of the late Professor Frederick Soddy, F.R.S., to assist local and regional studies are now prepared to consider inquiries from persons interested in such projects . . .

and went on to rehearse the terms of the Trust, that any grant should be to groups, preferably of students or teachers, studying the social, economic and cultural life of specific regions, along the general lines of the Le Play Society. However, a rider was added: 'the Trustees emphasize that they have wide powers of discretion.'[3]

The key elements of the conditions on which grants could be made were set out in the will as follows:[4]

2. To use from time to time the accruing income for the following general educational purposes namely for the development of study primarily in the area concerned but also by Group-Meetings held in homes or elsewhere of the social economic and cultural life of specific regions either in Great Britain and Ireland or in other countries anywhere else in the World along the general lines initiated developed and being now carried on in Great Britain under the able guidance if its Director by The Le Play Society . . .
3. In awarding Grants for such purposes it is my wish that preference should be given other things being equal (a) to members of the teaching profession and students of or those who have studied the sociological sciences (b) to the younger men and women among such teachers and students early in their careers (c) by those otherwise unable to afford the expenses of travel and (d) for the purposes of Group-Study such as are among the main present activities of The Le Play Society rather than to individuals alone.

However, it was also clear that it was the 'old' Le Play Society of the 1920s and 1930s, with its youthful energy and commitment to roughing it to find 'place–work–folk', rather than its more comfortable descendants, which Soddy wished to support.

The response to the announcement in the press 'was on such a scale and of such a nature as to set the Trust into operation without delay.'[5] At the first of the regular biennial meetings of the Trustees on 22 June 1957, the applications were examined and divided into four categories: the first included those applications which fell outside the terms of the Trust 'but revealed an extraordinary variety of things that groups of people in this Country wanted to do'; these were rejected summarily.[6] Secondly, there were those which were unsuitable because they were in the nature of hobbies rather than serious academic studies. Thirdly, there were applications from official or university departments; and finally those 'from private individuals proposing what appeared to be worthwhile undertakings within the Trustees' terms of reference'. Selection was made from the last group, and preference was given to group work, to young people, and to actual study and research in the field which would lead to a published report and especially to encourage expeditions from schools.[7] Thirty applications were rejected. By the time of the second Trustees' meeting in November 1957, the situation was becoming clearer. The financial position was resolved, and after all necessary duties and

EPILOGUE 199

expenses were paid there was a sum of £19 750 remaining and an annual income of £1000.

No further developments were possible until after the next meeting, except a visit to the Museum of English Rural Life at the University of Reading in February 1958, which led to a two-year grant 'for a detailed study of the economic, social and cultural developments in a Berkshire downland parish'.[8] The results of this survey, and other materials combined with it, were published in 1966 as *Estate Villages*, a report which has stood the test of time and still remains a key text in the area.[9] The trustees' judgement that this was 'the most notable of all the reports they received' has been amply borne out.[10] This particular grant is also a good example of one of the ways in which relatively small grants from the Trust have been used as 'pump primers'. More than one recipient of such an award has stressed the difficulty of raising the first sum of money. There is a general reluctance to provide this in any form at all, but once an initial sum has been awarded, subsequent grants are much easier to obtain.

At the meeting in May 1958, a further point of principle was agreed in considering the eligibility of projects. Other things being equal, 'special value should be attached to organised co-operation with local workers when groups go to undertake field studies in a foreign country.'[11] In addition, the Trust has always desired especially to encourage expeditions from schools.

The result of these careful formulations of policy and examination of all applications led to the first grants being made to eleven projects. The successful projects included grants for students to take part in four different Le Play study tours, in Corsica, Eire, Morocco and Germany. There were also visits by other university and school groups to destinations from England to the Pyrenees. The proposals for these successful applications illustrate the extent to which the Le Play objectives of folk/work/place were adhered to, whether in a Berkshire parish with its aim of a detailed study of the economic, social and cultural development, in Corby where the emphasis was on the relationship of the social structure to the local region or 'A case study of two Sri Lankan Villages'.[12]

After their first three years, in 1960, the members of the Trust paused to review the progress of their work. The Trustees concluded that they had progressed satisfactorily. The financial position was sound, the applications for grants continued and a number of 'very worth-while' projects had been completed. However, 'the Trustees came to the conclusion that more could be done . . . if something of a more permanent nature could be established under the wing of a University. From this, a decision was reached . . . to offer to the new University of Sussex a "Frederick Soddy Research Fellowship in Geographical Sociology".'[13] Sussex was chosen not only because it was near where Soddy had lived for the last years of his life, but because the 'new map of learning' which the University sought to create seemed particularly appropriate.

Consequently, a letter was sent to the Registrar in August 1962, offering £1000 a year for three years for a Research Fellow to develop regional studies along the lines set out by Professor L. Dudley Stamp and emphasising the importance of geography 'in keeping sociological studies based on all the realities of the environment'.[14] Discussions began with the Vice-Chancellor, J. S. (later Lord) Fulton, then with the Professor of History and Pro-Vice Chancellor, Professor Asa Briggs (now Lord Briggs of Lewes). They were able to report by October 1963, 'the Trustees and the University have worked in complete harmony in bringing this new departure into realisation'.[15] By 1962 arrangements were completed, and a selection committee consisting of Professor Briggs, Professor Elkins (Professor of Geography), Mr Bunker and Mr Shillan had appointed B. T. (Barry) Wood of the University College of Aberystwyth.

The new appointment was announced in the local press in October 1962: 'Research Fellowship for Sussex Varsity' and 'Geographical Sociology New Fellowship Founded'.[16] The university press release explained that 'the obvious choice for this [fellowship] was the University of Sussex, not only because it is situated in Brighton, the headquarters of the Trust, but because the way in which it proposes to develop academic work is admirably adapted to furthering the basic interests of the Trust and of Professor Soddy himself.'[17] The first fellow, Barry Wood, seemed particularly appropriate as he had come from Aberystwyth where Soddy himself had been a student in the 1890s and, as described above, had retained particularly pleasant memories of his time there.

The founding of the Fellowship marked the beginning of a long and fruitful relationship between the Trust and the university. The Fellowship was the first in the arts area of the new university, and it was established when the present buildings at Falmer were hardly begun, and while the fledgling establishment was housed, literally, in Preston Road in Brighton. In 1967, T. H. Elkins, Foundation Professor of Geography, remembered the first contacts, and also summed up the key importance of the trust in the University's founding years.[18]

Thinking back to Preston Road, when the Frederick Soddy Research Fellowship was the sole provision for social research in the whole university, it is gratifying to contemplate the position now, with a lively interdisciplinary group of faculty members and research workers occupied with what we now tend to call 'environmental studies' . . . All this is squarely in the Le Play/Geddes/Soddy tradition, and it might not have happened at all without the first 'pump priming' by the Frederick Soddy Trustees. We were really most fortunate to be singled out for this imaginative piece of research support.

The first Fellow, Barry Wood, showed the new university's commitment to its immediate neighbourhood by working on the social structure and problems of a 'new' town, Crawley. During his tenure a considerable amount of vital data was collected and a group of working papers produced on local social structures and the character of local relations with the community. Dr Wood's fellowship also had the vital role of cementing the relationship between the Trust and the University initially through his involvement in the work of the geography department and then by his appointment to the faculty at the end of his fellowship.

The close connection between the Trust and the university was continued by the second Fellow, Dr A. R. Fielding, who used his period as Soddy Fellow to develop his earlier studies on migration in France in a comparative and theoretical framework. This work resulted in publications on the ways in which regional economic growth impacted on interregional migration in France through the workings of urban and regional labour markets.[19] Tony Fielding also remembers that during his fellowship he spent a good deal of time developing courses—a vital job in a still new university. This illustrates the extent to which contributing to the development of the University of Sussex, and especially, in the early days, to geography, has been one of the numerous 'hidden extras' featuring in the work of successive Fellows. After his period as a research fellow, Tony Fielding was also appointed to the faculty, where he continues to teach.

This pattern could not be expected to continue. The three Fellows and one student from 1966 to 1973 completed valuable studies on a range of topics but never repeated the role within the University that Wood and Fielding had established. H. F. Andrews (1966–8) used his fellowship to complete his doctoral thesis which used a development of Geddes's place/work/folk triad to examine the way in which people are grouped around shopping facilities of various kinds. Another version of the established Geddes/Le Play theme was examined in the studies of L. Braithwaite (1968–9) who worked on the aesthetic aspects of the built environment. After this appointment the fellowship followed developments in geography into more quantitative studies. However the appointment of Fred Gray (1974–7) marked a return to the more 'humanistic' concerns of the earlier period as his reports on local authority housing show. Fred Gray also revived the earlier tradition through his appointment as lecturer in the School of Cultural and Community Studies at the end of his fellowship.

From the early days, there have been receptions held by the Trustees for members of expeditions to report on their activities, to celebrate the appointment of new Fellows and to entertain distinguished guests and friends of the Trust. The first of these was held at the London School of Economics and they were then transferred to the Royal Geographical Society, where they continue to take place. More recent meetings have acquired something of the character of those of the Le Play Society winter conferences in Oxford. Representatives of at least some of the projects funded by the Trust present verbal, often illustrated, accounts of their work. Afterwards, there is normally an informal gathering which gives the various groups an opportunity to discuss their successes and difficulties.

Links between the Trust and geographers at the University of Sussex and with the Royal Geographical Society have drawn closer. Three of the original Trustees have now retired. Margaret Tatton was the first and she died soon afterwards. Then Marjorie Soddy was afflicted with failing eyesight and so could no longer continue her duties, although she still takes a keen interest in the Trust from the nursing home to which she moved. David Shillan died suddenly and is much missed. One of the new Trustees, Dr Helen Wallis, formerly Keeper of the Map Room of the British Museum and from 1973 Map Librarian of the British Library, died in 1995. The others are W. R. Mead, Emeritus Professor of Geography at University College, London, and David Hall, some time Honorary Foreign Secretary of the Royal Geographical Society. All members of the Trust are Fellows of the Royal Geographical Society, as is the chairman, despite his major qualification being School Certificate geography. At one of the receptions at the Royal Geographical Society, Lord Hunt, of Everest fame, the then President, suggested to Peter Bunker that he should become a Fellow in order to ensure the close association of the two organisations, and so the non-geographer has the privilege of FRGS after his name.

The reception is not the only Le Play influence on the Trust-funded projects. Members of expeditions of all kinds are required to write a report on their activities and to provide the Trustees with a copy. Since each one is unique only a representative sample can be mentioned here. They range from professional publications to typescripts. For the research on his professionally published *An Agricultural Atlas of Scotland*, T. J. Coppock, Ogilvie Professor of Geography in Edinburgh, had funds from the Carnegie, Leverhulme and MacRobert Trusts, as well as from the Frederick Soddy Trust. *The Royal Geographical Society Oman Sands Project*, an academic study undertaken by a large team of researchers under the auspices of the Royal Geographical Society, had a three page list of sponsors and was published as a highly professional report.

In complete contrast, there are reports from school expeditions. The Evesham High School Expedition, for example, went to Iceland. Ten boys and three members of staff produced a typewritten account, illustrated with photographs and hand-drawn sketches, which, although it does not contain material suitable for publication, shows how much they learnt about Icelandic society and about working together as a team. A similar kind of experience is reflected in the report of the Geography Department of the Queen Elizabeth Grammar School in Darlington, the students of which carried out 'A Geographical Survey of Middle Swaledale'.

There are, between these extremes in terms of funding and of presentation, a number of reports from university departments and student groups which represent more conventional academic studies. The Geographical Field Group, founded as a younger group following in the Le Play tradition, received more than two dozen grants from the Trust. Their reports ranged from studies undertaken around the North Atlantic fringe to Eastern Europe and West Africa. One of the most successful projects, in terms of published materials, was the study of 'Periodic and Daily Markets in Highland Ecuador' undertaken by R. J. Bromley, assisted by his wife. Bromley not only completed his D.Phil thesis but then published six academic articles from his research.

The general flavour of the Trust's grants can, perhaps, best be appreciated through a description of the reports from one year. In 1988, eight reports were received. Five of the expeditions were by students from various university departments or societies, one from an exploration group, one from a local Community Project and one from a school project. Two groups visited China, the others went to Jamaica, Brazil, Borneo, and Pakistan, while one remained in the local community. The size of groups ranged from the three social anthropologists who visited Jamaica to complete field work for their MA dissertations to the 62 members of the Yorkshire Schools Exploring Society who undertook field studies in China.

The 44 young people and 18 adult leaders were divided into five groups, one remaining at the base camp, the others exploring the region by canoe, mountaineering, mountain bike or trekking. The organisation which this necessitated was remarkable. From the initial advertising, through selection, and activity weekends for training, total commitment was required by all concerned. What stands out in the report submitted to the Trust is the sheer hard work, imagination and initiative shown by all of the young people. They were required to bear the greater part of the cost themselves, and this meant finding £1950 each. Local businesses were contacted, sponsored activities organised, stalls at car boot sales set up and a number of other fund raising methods undertaken. Training weekends were needed for physical fitness and learning necessary skills such as map reading, cooking and setting up programmes of activity. Food requirements had to be calculated, Chinese authorities had to be dealt with, clothing requirements decided. Finally, the group set off—by coach to London, plane to Beijing, train to Xian and Xening and finally on to the easten side of the Tibetan plateau. It was a privilege to attend the meeting of the Trust at which the group leaders reported back. The accounts of trying to keep track of lively teenagers on mountain bikes with varied levels of expertise, or the problems the canoeing group encountered with temperatures rising to 90 Fahrenheit during the day to be followed by frost at night, will always be remembered. Most impressive was the achievement of taking two handicapped children. There were problems. To quote: 'A couple of times the walking was difficult; once we came to a 100ft. cliff and had to descend to the river—that was pretty scary because it was loose ground—shingle. I felt a bit unsure because you never knew where you were putting your feet'. It was 'scary' because the writer was blind, relying on the assistance of other members to tell him what was happening.

A contrast again is to be found in Peter Ambrose's study of the Sussex village of Ringmer which was completed with assistance from the Trust. Dr Ambrose teaches Urban Studies at the University of Sussex and so through this publication the links between the trust, the university and the local community were continued.[23]

The quality and number of the reports, now over 200, fully justify the confidence shown by Soddy in his Trustees when he gave them 'wide powers of discretion' in his will. Some of the projects could not have taken place without the Trust, which was the sole funding body. In other cases the Trust played a helpful pump-priming role.

All the reports are deposited in the Library of the University of Sussex, where they are accessible and available in the public domain. Collectively they contain a mass of information, both specifically about particular places and people, and generally on matters such as funding and equipping expeditions. The accounts of fund-raising alone deserve a book.

The funding of the smaller projects and of the fellowship continue side by side but there have been changes in the arrangement for the fellowship. In the first place, consequent upon changes in the university structure, the fellowship has been transferred within the university from the Department of Geography to the department of 'Urban and Regional Studies' which later became 'The Centre for Urban and Regional Research'. Secondly, since the remuneration of the fellowship had always been linked to national salary scales and substantial increases took place in the late 1970s, increase in the demands on the Trust was experienced. A temporary solution was found through joint funding of the fellowship, with a contribution from the National Council of Social Service when a project on the welfare provision of the churches in a parish setting was supported.[21] Although studies from the Church Project were completed and a report was prepared for the Lewis Cohen Urban Studies Centre in Brighton,[22] the informal arrangement proved unsatisfactory. As a result, the terms of the appointment of the subsequent holder of the fellowship, Beryl Day, were formalised more precisely with the university and specifically with the Centre for Urban and Regional Research. The arrangement proved satisfactory. Ms Day's MPhil on the local community in Hove was published in 1992 as *Dependency; Personal and Social Relations*.[23]

Finally, this biography is also the result of a Frederick Soddy Research Fellowship. It has perhaps stretched the terms of the Trust somewhat further than Soddy might have envisaged, but has been in the minds of the Trustees since at least 1973. A surprising number of people from historians of science, and those who knew him, especially in the inter-war period, to those who have benefited from his will have felt that such a review of the man and his times was needed. It is hoped that the story does Frederick Soddy justice and explains how geography and the social sciences came to benefit from the testament of a Nobel Laureate in Chemistry.

Notes and references

1. This comes from the text of a speech given by Shillan in 1982 which is to be found in the papers of the New Europe Group in the New Atlantis Archive.
2. This material comes from interviews with Peter Bunker, Marjorie Soddy, and Margaret Lewis, and from the text of Peter Bunker's speech at the reception of the trustees for guests and members of the Royal Geopgraphical Society on 19 January 1988.
3. *The Times*, 22 May 1957, p. 8; *The Times Educational Supplement*, 26 May 1957; *Manchester Guardian*, 29 May 1957.
4. Information from the Soddy trustees.
5. *The Report of the Frederick Soddy Trust*, September 1958, p. 5. (This is the First Report.)
6. Peter Bunker's speech, 19 January 1988.
7. Peter Bunker's speech, 19 January 1988, p. 6.
8. *Report of the Frederick Soddy Trust*, September 1958, p. 15.
9. Havinden, M. (1966). *Estate Villages*. Lund Humphries.
10. *Third Report of the Frederick Soddy Trust*, November 1967, p. 4.
11. *Report*, 1958, p. 8.
12. These details all come from the First Report.
13. *The Second Report of the Frederick Soddy Trust*, October 1963, p. 4.
14. Letter from the trustees to A. E. Shields, Registrar. Copy in Frederick Soddy files.
15. *Second Report*, p. 4.
16. *The Evening Argus*, 23 October 1962; *The Brighton and Hove Herald*, 27 October 1962.
17. Copy of this in the Frederick Soddy files held in the University of Sussex.
18. Letter of Professor T. H. Elkins to P. J. Bunker. Soddy Files, CCS, University of Sussex, dated 13 November 1967.

19. Fielding, A. J. *Urban Studies.* November 1966.
20. Ambrose, P. (1974). *The Quiet Revolution. Change in a Sussex Village 1870–1970.* Chatto and Windus.
21. Material about this from files in CCS, University of Sussex.
22. For example, Beresford, P. and Croft, S. (1986). *Whose Welfare: Private Care or Public Services.* The Lewis Cohen Urban Studies Centre, Brighton.
23. Avebury Press, Aldershot.

Bibliography

Manuscript sources

Aberdeen University Library. Department of Special Collections.
Bodleian Library, Oxford. Soddy Papers; Oxford University Archive.
Cambridge University Library. Rutherford Collection.
East Sussex Record Office. Census; Church Records.
Fawcett Library, London. Arncliffe–Sennett Papers.
McGill University, Montreal. Archives.
Mitchell Library, Glasgow. Suffrage papers.
Museum of the History of Science, Oxford. Additional MSS.
New Atlantis, Ditchling. Archive and papers.
Royal Society. Council Documents; Personal Records; J. Larmor MSS.
Royal Institution. WIlliam Henry Bragg Collection.
University College, London. Lodge MSS; Travers MSS.
University of Bristol, University Library. Beilby Papers.
University of Glasgow Archives. University Court Books.
University of Illinois Library at Urbana-Champaign Rare Books Room. Wells Collection.
University of Keele Library. The Branford (Le Play Society) Papers; Papers of the Sociological Society; S. H. Beaver MSS, uncatalogued.
University of Strathclyde. Beilby Papers; John Arnold Cranston Papers; Geddes Papers.
University of Sussex Library. Papers of The Frederick Soddy Trust; Reckitt Papers.

Interviews (unless otherwise stated with the author)

Peter Bunker, Hove, Sussex, 1995.
Margaret Lewis, London, 1991.
Harry Rutherford, Ditchling, Sussex, 1990–1.
Gracie Rutherford, Ditchling, Sussex, 1990–1.
Maragaret Shillan, Ditchling, Sussex, 1991.
Marjorie Soddy, Oxford, 1990.
Ralph Twentyman, Forest Row, Sussex, 1991.
Reginald Wrugh, London (with Stuart Laing), 1989; with the author, 1990–1.

Newspapers and periodicals

Aberdeen Free Press
Aberdeen University Labour Club Magazine
Alma Mater (Aberdeen)
Calvacade
Credit Power
Daily Herald
Forward
The Guildsman
Nature
New Age
New Atlantis
New Britain
Public Welfare
Scientific Worker
The Times (London)

Frederick Soddy's works

This list contains the most important of Soddy's works. There are a number of minor, jointly written articles which have not been listed here. A complete listing of Soddy's scientific writings may be found in M. Howorth, *Pioneer Research on the Atom*.

1894	with R. E. Hughes, 'The Action of Dried Ammonia on Dried Carbon Dioxide Gas', *Chemical News*, 22 March 1894.
1898	'The Life and Work of Victor Meyer', *Transactions O.U.J.S.C.*
1902	with Rutherford, 'The Cause and Nature of Radioactivity', Parts I and II, *Philosophical Magazine*, **4**, 370–96 and 569–85.
1902	with Rutherford, 'The Radioactivity of Thorium Compounds', Parts I and II, *Transactions of the Chemical Society*, **81**, 321–50 and 837–51.
1902	'The Radioactivity of Uranium', *Transactions of the Chemical Society*, **81**, 860 ff.
1903	with Rutherford, 'Radioactive Change', *Philosophical Magazine*, **6**, 576–91.
1903	with Rutherford, 'A Comparative Study of the Radioactivity of Radium and Thorium', *Philosophical Magazine*, **6**, 445–57.

1903	with Rutherford, 'Condensation of the Radioactive Emanations', *Philosophical Magazine*, **6**, 561–76.
1903	with Rutherford, 'Radioactivity of Uranium', *Philosophical Magazine*, **6**, 441–5.
1903	with Ramsay, 'Experiments in Radioactivity and the Production of Helium from Radium', *Proceedings of the Royal Society*, **72**, 204 ff.
1903	with Ramsay, 'Further Experiments on the Production of Helium from Radium', *Proceedings of the Royal Society*, **72**, 346 ff.
1903	'The Disintegration Theory of Radioactivity', *The Times Literary Supplement*, 26 June 1903, 201.
1903	'Possible Future Applications for Radium', *The Times Literary Supplement*, 17 July 1903, 225 ff.
1903	'Some Recent Advances in Radioactivity', *Contemporary Review*, May 1903, 708–20.
1903	'Radium', *Professional Papers of the Royal Corps of Engineers*, **29**, 237–51.
1904	*Radio-Activity: an Elementary Treatise from the Standpoint of the Disintegration Theory*, The Electrician Publishing Co. Appearing also as articles in *The Electrician*, **52**, 1903, et seq.
1904–20	Series of annual reports on radioactivity for the *Annual Reports on the Progress of Chemistry*, since published as *Radioactivity and Atomic Theory*, London, 1975, edited and with a commentary by T. J. Trenn.
1904	The Wilde Lecture VIII, 'The Evolution of Matter as Revealed by the Radioactive Elements', 16 March 1904, in *Memoirs and Proceedings of the Manchester Literary and Philosophic Society*, **48**.
1905	'The Production of Radium from Uranium', *Philosophical Magazine*, **9**, 768.
1906	'The Evolution of the Elements', Report of the 76th meeting of the BAAS, pp. 122–31.
1906	'The Internal Energy of the Elements', *Journal of the Institution of Electrical Engineers*, **37**, 140–7.
1906	'The Present Position of Radioactivity', *Journal of the Rontgen Society*, **2**, 45–65.
1907	with T. D. Mackenzie, 'The Relation between Uranium and Radium', *Philosophical Magazine*, **14**, 272–80.
1907	'Calcium as as Absorbent of Gases for the Production of High Vacua and Spectroscopic Research', *Proceedings of the Royal Society*, **78A**, 429 ff.
1907	with R. von Hirsch, 'A Gas Generated by Aluminium Electrodes', *Philosophical Magazine*, **14**, 779–83.

1908	*Ion: A Journal of Electronics, Atomistics, Ionology, Radioactivity and Raumchemistry*. Soddy was the editor of this journal.
1908	'Le Radium', *Physikalische Zeitschrift*, **5**, 361–3.
1908	'The Founder of Radioactivity', obituary of Becquerel in *Ion* **1**, 2–4.
1908	'The Charge Carried by the α-particle', *Ion*, **1**, 4.
1908	'Attempts to Detect the Production of Helium from the Primary Radioelements', *Philosophical Magazine*, **16**, 573–5.
1908	'Die Wehnelt-Kathode im hochgradigen Vakuum', *Physikalische Zeitschrift*, **9**, 8.
1908–10	'Relation between Uranium and Radium,' Parts III, IV, V, *Philosophical Magazine*, **16**, 632 ff; **18**, 846 ff; **20**, 340 ff.
1908	with T. D. Mackenzie, 'The Electric Discharge in Monatomic Gases', *Proceedings of the Royal Society*, A **80**, 92.
1908	'Radium', *Transactions of the Glasgow Technical College Scientific Society*, 14 November.
1909	with Russell, 'The γ-rays of uranium and radium', *Proceedings of the Royal Society* **18**, 620 ff.
1909	'Multiple Atomic Disintegration', *Philosophical Magazine*, **18**, 739–40.
1909	*The Interpretation of Radium: Being the substance of six free experimental lectures delivered at the University of Glasgow, 1908.* John Murray.
1909	'Die Bildung von Helium aus Uranium', *Physikalische Zeitschrift*, **10**, 41; also in *Le Radium*, 1908, **5**, 361.
1909	'The Energy of Radium', *Harper's Monthly*, **120**, Dec. 1909, 52–59.
1909–10	'The Rays and Product of Uranium X', I and II, *Philosophical Magazine*, **18**, 858 ff and **20**, 42 ff.
1910–11	with Miss R. Pirrett, 'The Ratio between Uranium and Radium in Minerals', I and II, *Philosophical Magazine*, **20**, 345 ff and **21**, 652 ff.
1910	'Essais pour evaluer le periode de l'ionium', *Le Radium*, **7**, 295 ff.
1910–11	with A. J. Berry, 'The Conduction of Heat through Rarefied Gases', *Proceedings of the Royal Society*, A **83**, 254 ff and 576 ff.
1910	with Russell, 'The Constant of Uranium X', *Philosophical Magazine*, **19**, 847 ff.
1910	with Mrs W. M. Soddy and Russell, 'The Question of the Homogeneity of the γ-rays', *Philosophical Magazine*, **19**, 725 ff.
1910	Translation of *Brownian Movements and Molecular Reality*, by Jean Perrin, Taylor and Francis.
1911	*The Chemistry of the Radio-Elements*, Part 1, Longman, Green

	and Co. Leipzig, J. A. Barth, 1912; Part 2, Longman, Green and Co. 1914.
1911	'The Chemistry of Mesothorium', *Transactions of the Chemical Society*, **99**, 72–83.
1911	with Russell, 'The γ-rays of Uranium and Radium', *Philosophical Magazine*, **21**, 130 ff.
1912	*Matter and Energy*, Williams and Norgate.
1912	'Transmutation: The Vital Problem of the Future', *Scientia*, **11**, 186–202.
1913	'The Radio-elements and the Periodic Law,' *Chemical News*, **107**, 97–9. *British Association Report* and *Annual Report* contain slightly fuller versions of the paper.
1913	'Intra-atomic Charge', *Nature*, **92**, 399–400.
1913	*Abstracts of Chemical Papers: Journal of the Chemical Society* has Soddy's abstracts from papers by Russell, Fajans and Soddy.
1913	'The Origin of Actinium', *Nature*, **91**, 634–635.
1913	'Science on the Road to Revolutionise all Existence', *New York Times*, Magazine Section, Part 6, September 28, p. 6.
1914	with H. Hyman, 'The Atomic Weight of Lead from Ceylon Thorite', *Transactions of the Chemical Society*, **105**, 1402–8.
1914	'The Existence of Uranium Y', *Philosophical Magazine*, **37**, 215 ff.
1914	'Science and Life', *Candid Quarterly*, February 1914, 237–60.
1914	'The Nature of the Argon Family of Gases', *Science Progress*, April 1914.
1915	'The Density of Lead from Ceylon Thorite', *Nature*, **94**, 615.
1915	'The Social Effects of Recent Advances in Physical Science', lecture to Aberdeen WEA.
1915	'Some Aspects of the Atomic Theory', review of Bragg's book, *Science Progress*, July 1915.
1916	'Ramsay', obituary, *Nature*, **97**, 482.
1917	'The Separation of Isotopes', *Journal of the American Chemical Society*, **39**, 1614–19.
1917	'A Criticism of the Financial Operations of the Carnegie Trust for the Universities of Scotland', *Science Progress*, January 1917.
1918	*Scientific Research*, Labour Representation Committee for Scottish Universities.
1918	with Cranston and Miss Hitchins, 'The Parent of Actinium', *Proceedings of the Royal Society*, A **94**, 384 ff.
1918	'The Conception of the Chemical Element as Enlarged by the Study of Radioactive Change', *Transactions of the Chemical Society*, **115**, 211–16.
1918	Review of Margaret Todd, *The Life of Sophia Jex-Blake*, *Nature*, **101**, 461–2.

1920   *Science and Life: Aberdeen Addresses*, John Murray.
1920   'Economic "Science" from the Standpoint of Science', *The Guildsman*, No. 43, July 1920, 3 ff.
1920   'The Public Support of Scientific Work', *Scientific Worker*, Supplement to Vol. 1, No. 3, 15–20.
1921   'Contribution to Discussion on Isotopes', *Proceedings of the Royal Society*, A **99**, 97.
1922   The Nobel Lecture, 'The Origin of the Concept of Isotopes,' in *Les Prix Nobel en 1921–1922*, Stockholm, 1923.
1922   'Isotopes', introduction to discussion on 'Atomic Structure', Reunion de Bruxelles, April 1922.
1922   *Cartesian Economics: the bearing of physical science on State stewardship*, two lectures to the Student Union, Birkbeck College and London School of Economics, 10 and 17 November 1921, Henderson's, May and October 1922 and March 1924.
1923   'Conrad Rontgen', obituary, Glasgow Herald, February 1923.
1924   *The Inversion of Science and a scheme of Scientific Reformation*, based on lectures given in 1923, Henderson's, 1924 (two editions) and 1927.
1926   *Wealth, Virtual Wealth and Debt: The Solution of the Economic Paradox*, George Allen and Unwin Ltd.
1927   *The Wrecking of a Scientific Age*, Henderson's.
1928   *The Impact of Science upon an Old Civilisation*, Henderson's.
1928   'What I think of socialism', *Socialist Review*, August 1928, pp. 28–30.
1931   *Money versus Man*, Elkin Matthews and Marrott Ltd., later editions George Allen and Unwin Ltd.
1932   *The Interpretation of the Atom*, Murray.
1932   *Poverty Old and New*, lecture to the New Europe Group, London, published by The Search Publishing Co. Ltd.
1932   'The α-rays of Ionium', *Nature*, **130**, 364.
1933   *Wealth, Virtual Wealth and Debt*, George Allen and Unwin Ltd. 2nd edn.
1933   'John Watts', obituary, *Journal of the Chemical Society*.
1933   'Reminiscences of McGill, 1900–1902', *Old McGill* annual, **1936**, 1933.
1934   'A Physical Theory of Money', paper to the Liverpool Engineering Society, *Transactions of the Liverpool Engineering Society*, **56**.
1934   *The Role of Money*, G. Routledge and Sons Ltd.
1934   'The role of money', *The Oxford Magazine*, June 7.
1934   'The New Britain Movement', *Supplement to New Britain*, June 20. Soddy was a regular contributor to *New Britain Weekly*, *New Britain Quarterly* and *The Eleventh Hour* throughout 1933 and 1934.

1934	'Contribution to discussion on heavy hydrogen', *Proceedings of the Royal Society*, A **144**, 11.
1935	'Money as Nothing for Something', 'The Gold Standard Snare', reprinted from *Garvin's Gazette*, Gibbs and Banforth, St. Albans.
1935	Foreword to *The Frustration of Science*, (ed.) Sir Daniel Hall, G. Allen and Unwin.
1936	'The pound for pound system' in Butchart, M. (ed.) *To-morrow's Money*, Stanley Nott.
1936	'Is Science a Failure', *Radiography*, July 2.
1936	'The Kiss Precise', *Nature*, **137**, 1021.
1936	'The Hexlet', *Nature*, **138**, 958.
1937	'Rutherford', obituary in *Nature*, **101**, 461–462.
1937	*Credit, Usury, Capital, Christianity, and Chameleons*, the Economic Reform Club.
1937	'The Bowl of Integers and the Hexlet', *Nature*, **139**, 77–9, 154.
1938	*The Budget, synopsis in one hundred verses of the author's 'Reformed Scientific National Monetary System'*, published by the author, Knapp, Enstone, Oxon.
1938	*Money and the Constitution*, published by the author, Knapp, Enstone, Oxon.
1938	'Social Relations of Science', *Nature*, **141**, 784–5.
1940	'Finance and the War', *Nature*, **147**, 449.
1941	*A Question the Brains Trust did not Answer*, Economic Reform Club.
1941	'Qui s'accuse s'aquitte', review of Hardy's *A Mathematician's Apology*, *Nature*, **147**, 3.
1942	'The Summation of Infinite Harmonic Series', *Proceedings of the Royal Society*, A **179**, 377.
1943	*The Arch-Enemy of Economic Freedom*, published by the author, Knapp, Enstone, Oxon, 1943.
1943	'Newton and Leibniz and p.', February 1943, 'Science for Rulers', November 1943, Presidential Address to the Science Master's Association, *The School Science Review*, **95**.
1943	'The Three Infinite Harmonic Series and their Sums', *Proceedings of the Royal Society*, A **182**, 113.
1945	'The Moving Finger Writes . . .', *Cavalcade*, 18 August 1945, 8–9.
1945	'The Atomic Age', *The Scottish Field*, October 1945, 225.
1947	'The Evil Genius of the Modern World', *Management and Human Relations in Industry*, Vol. 1, Industrial Relations Publishing Corporation, New York.
1947	'An independent scientist's views on the economic and political possibilities of Atomic Energy for the Future', address to the British Constitutional Research Association.

1948	'Can a Nation be Saved by Budgets and Ever Increasing Taxation', *The Torch of Truth*.
1949	*The Story of Atomic Energy*, Nova Atlantis.
1950	'Frederick Soddy Calling All Tax Payers', Economic Research Council.
1950	*Money Reform as the Preliminary to all Reform*, address to the Birmingham Paint Varnish and Lacquer Club, published by the Club.
1950	*Dishonest money or why a larger pay-packet now buys less than it did*, New Boadicea Club.
1951	*The Constitutional Justification for Resisting Tax-payments*, Economic Reform Club and Institute.
1951	'How Isotopes were Discovered', BBC talk for Social Survey series of science programmes.
1953	'Just Fifty Years Ago', address to the Institute of Atomic Information for the Layman, 28 February 1953 in Muriel Howorth, *Atomic Transmutation The Greatest Discovery Ever Made*, New World Publications.
1953	*Isotopes*, John Murray. *The Story of Atomic Energy*, 2nd edn., New World Publications.
1954	*Isotopy, Lectures by Soddy and Cranston*, New World Publications, 1954.
1956	*The Cubic Equation and a Machine that Solves it*, New World Publications.

Books (including printed primary sources)

Abrams, P. (1968). *The Origins of British Sociology: 1834–1914*. The University of Chicago Press.

Aldcroft, D. H. (1970). *The Inter-War Economy: Britain, 1919–1939*. Batsford.

Anderson, R .D. (1988). *The Student Community at Aberdeen 1860–1939*. Aberdeen University Press.

Anon. (1923). *Les Prix Nobel en 1921–1922*. Stockholm. (This is the original version of *Nobel Lectures*.)

Badash, L. (ed.) (1969). *Rutherford and Boltwood: Letters on Radioactivity*. Yale University Press, New Haven.

Barnes, B. (1985). *About Science*. Basil Blackwell.

Blythe, R. (1964). *The Age of Illusion England in the Twenties and Thirties 1919–1940*. Penguin Books, Harmondsworth.

Boardman, P. (1978). *The Worlds of Patrick Geddes*. Routledge and Kegan Paul.

Bramwell, A. (1989). *Ecology in the Twentieth Century: A History*. Yale University Press, New Haven.
Branson, N. (1975). *Britain in the Nineteen-Twenties*. Weidenfeld and Nicolson.
Bunge, M. and Shea, W. (ed.) (1979). *Rutherford and Physics at the Turn of the Century*. Dawson, New York.
Burgess, K. (1980). *The Challenge of Labour: Shaping British Society 1850–1930*. Croom Helm.
Cannadine, D. (1980). *Lords and Ladies: the Aristocracy and the Towns 1774–1967*. Leicester University Press.
Chadwick, J. (ed.) (1962). *Collected Papers of Lord Rutherford of Nelson*, Vol. 1. Allen and Unwin.
Chant, C. and Fanvel, J. (ed.) (1980). *Darwin to Einstein: Historical Studies of Science and Belief*. Longman, Harlow.
Clark, I. F. (1992). *Voices Prophesying War: Future Wars 1763–3749*. Oxford University Press.
Colby, R. C. et al. (1990). *Companion to the History of Modern Science*. Routledge.
Cole, G. D. F. (ed.) (1933). *What Everybody Wants To Know About Money: A planned outline of monetary problems*. Victor Gollancz.
Cole, Dame Margaret (1971). *The Life of G.D.H. Cole*. Macmillan.
Crosbie-Smith, D. (1989). *Energy and Empire: A Biographical Study of Lord Kelvin*. Cambridge University Press.
Delahaye, J. V. (1929). *Politics: A Discussion of Realities*. Initiated by J. V. Delahaye, in company with Hilderic Cousens, V. A. Demant, Philippe [sic] Mairet, Albert Newsome, Alan Porter, Maurice B. Reckitt, W. T. Symons. The C. W. Daniel Company.
Dickson, L. (1969). *H.G. Wells: His Turbulent Life and Times*. Macmillan.
Easlea, B. (1983). *Fathering the Unthinkable: Masculinity, Scientists and the Nuclear Arms Race*. Pluto Press.
Eve, A. S. (1939). *Rutherford, being the Life and Letters of the Rt. Hon. Lord Rutherford, O.M.* Cambridge University Press.
Farnell, L. R. (1934). *An Oxonian Looks Back*. Martin Hopkinson Ltd.
Feather, N. (1973). *Lord Rutherford*. (1st edn. 1940.) Priory Press.
Finlay, J. L. (1972). *Social Credit: The English Origins*. Queen's University Press, McGill.
Fussel, P. (1975). *The Great War and Modern Memory*. Oxford University Press.
Gardner, M. (1979). *Mathematical Circus*. Penguin, Harmondsworth.
Gibbon, L. G. and MacDiarmid, H. (1934). *Scottish Scene, or The Intelligent Man's Guide to Albyn*. Jarrolds.
Gillispie, C. G. (1975). *Dictionary of Scientific Biography*. Charles Scribner's Sons, New York.
Gowing, M. (1964). *Britain and Atomic Energy 1939–1945*. Macmillan.
Gowing, M., assisted by Arnold, L. (1974). *Independence and Deterrence. Britain and Atomic Energy, 1945–52. Volume 2. Policy Execution*. Macmillan.
Green, V. H. H. (1974). *A History of Oxford University*. Batsford.
Griffiths, R. (1983). *Fellow Travellers of the Right. British Enthusiasts for Nazi Germany 1933–39*. Oxford University Press.

Haber, L. F. (1986). *The Poisonous Cloud: Chemical Warfare in the First World War*. Clarendon Press, Oxford.
Haberler, G. (1963). *Prosperity and Depression*. Atheneum, New York.
Hackmann, W. (1984). *Seek and Strike: Sonar, Anti-submarine Warfare and the Royal Navy 1914–1954*. HMSO.
Hall, Sir D. et al. (1935). *The Frustration of Science*. George Allen and Unwin.
Hardy, G. H. (1940). *A Mathematician's Apology*. Cambridge University Press.
Harrison, B. (ed.) (1994). *A History of the University of Oxford*. Vol. 8. *Twentieth Century Oxford*. Oxford University Press.
Harrison, T. (1992). *Square Rounds*, Faber and Faber.
Harrod, R. F. (1959). *The Prof. A Personal Memoir of Lord Cherwell*. Macmillan.
Holloway, E. (1986). *Money Matters: A Modern Pilgrim's Economic Progress*. The Sherwood Press.
Holton, H. H. (1986). *Feminism and Democracy: Women's Suffrage and Reform: Politics in Britain 1900–1918*. Cambridge University Press.
Howorth, M. (1953). *Atomic Transmutation: The Greatest Discovery Ever Made from Memoirs of Frederick Soddy, M.A., LL.D., F.R.S., Nobel Laureate 1921*, Vol. 1. New World Publications. (There were no further volumes.)
Howorth, M. (1958). *Pioneer Research on the Atom: Rutherford and Soddy in a glorious chapter of science: The Life Story of Frederick Soddy M.A., LL.D., F.R.S., Nobel Laureate*. New World Publications.
Howorth, M. *Science and Philosophy: Extracts from the Works of Frederick Soddy*. (No publisher, no date, but contains a 1956 photograph of Soddy.)
Howorth, M. (1960). *Atomic Gardening for the Layman*. New World Publications.
Kauffman, G. B. (ed.) (1986). *Frederick Soddy (1877–1956): Early Pioneer in Radiochemistry*. D. Reidel Publishing Company, Dordrecht.
Levitas, R. (1985). *The Ideology of the New Right*. Polity, Cambridge.
Lunn, K. and Thurlow, R. (ed.) (1980). *British Fascism: Essays on the Radical Right In Inter-War Britain*. Croom Helm.
MacCarthy, F. (1989). *Eric Gill*. Faber and Faber.
Mairet, P. (1936). *A.R. Orage: A Memoir*. J. M. Dent and Sons.
Mairet, P. (1981). *Autobiographical and Other Papers*, (ed. C. H. Sisson). Carcanet, Manchester.
Martin, W. (1967). *The New Age under Orage*. Manchester University Press.
Martinez-Alier, J. with Schlupman, K. (1987). *Ecological Economics: Energy, Environment and Society*. Basil Blackwell, Oxford.
Marwick, A. (1991). *The Deluge: British Society and the First World War*. Macmillan.
Maxwell, D. W. (1932). *The Principal Cause of Unemployment*. Williams and Norgate.
Meller, H. (1990). *Patrick Geddes: Social Evolutionist and City Planner*. Routledge.
Middlemas, R. K. (1965). *The Clydesiders*. Hutchinson.
Mirowski, P. (1989). *More Heat than Light. Economics as Social Physics: Physics as Nature's Economics*. Cambridge University Press.
Moggridge, D. E. (1992). *Maynard Keynes: An Economist's Biography*. Routledge.
Morris, M. et al. (1976). *The General Strike*. Penguin Books, Harmondsworth.
Morris, W. (1977). *News From Nowhere*. Routledge Keegan and Paul.

Muir, E. (1940). *The Story and the Fable: An Autobiography*. George Harrap and Co. Ltd.
Muir, W. (1968). *Belonging: A Memoir*. The Hogarth Press.
Mumford, L. (1979). *My Works and Days: A Personal Chronicle*. Harcourt Brace Jovanovich.
Myers, M. G. (1940). *Monetary Proposals for Social Reform*. Columbia University Press, New York.
Newman, M. (1989). *John Strachey*. Manchester University Press.
Orage, A. R. (1917). *An History of Economics*. Fisher Unwin.
Peart-Binns, J. (1988). *Maurice Reckitt: A Life*. The Bowerdean Press and Marshall Pickering, Basingstoke.
Perkins, H. (1989). *The Rise of Professional Society: England since 1880*. Routledge.
Pimlott, B. (1977). *Labour and the Left in the 1930s*. Cambridge University Press.
Poole, J. B. and Andrews, K. (ed.) (1972). *The Government of Science in Britain*. Weidenfield and Nicolson.
Porter, A. (ed.) (1927). *Coal: A Challenge to the National Conscience*. Hogarth Press.
Redman, T. (1991). *Ezra Pound and Italian Fascism*. Cambridge University Press.
Rhodes, R. (1986). *The Making of the Atomic Bomb*. Penguin, Harmondsworth.
Rigby, A. (1984). *Initiation and Initiative: An Exploration of the Life and Ideas of Dimitrije Mitrinovic*. Columbia University Press, New York.
Roderick, G. W. and Stephens, M. D. (1972). *Scientific and Technical Education in Nineteenth Century England*. David and Charles, Newton Abbot.
Rose, H. and Rose, S. (1969). *Science and Society*. Allen Lane.
Ruskin, J. (1968). *Unto This Last*. Everyman Edition.
Russell, A. S. (1961). *Great Chemists*, (ed. E. Farber). Interscience Publishers.
Rutherford, H. C. (ed.) (1987). *Certainly, Future: Selected Writings by Dimitrije Mitrinovic*. Columbia University Press.
Selver, P. (1959). *Orage and the New Age Circle*. George Allen and Unwin.
Smith, D. C. (1986). *H.G. Wells: Desperately Mortal: A Biography*. Yale University Press, New Haven.
Smith, F. W. (The Earl of Birkenhead) (1961). *The Prof in Two Worlds*. Collins.
Smout, T. C. (1986). *A Century of the Scottish People 1830–1950*. Collins.
Snow, C. P. (1961). *Science and Government*. Oxford University Press.
Stalley, M. (1972). *Patrick Geddes: Spokesman for Man and the Environment*. Rutgers University Press, New Brunswick.
Taylor, A. J. P. (1970). *English History 1914–1945*. Penguin Books, Harmondsworth.
Todd, M. (1918). *The Life of Sophia Jex-Blake*. Macmillan.
Travers, M. W. (1956). *A Life of Sir William Ramsay*. Edward Arnold.
Trenn, T. J. (1977). *The Self-Splitting Atom: The History of the Rutherford–Soddy Collaboration*. Taylor and Francis Ltd.
Varcoe, I. (1974). *Organising for Science in Britain: A Case Study*. Oxford University Press.
Walker, G. (1988). *Thomas Johnson*. Manchester University Press.
Watts, A (1973). *In My Own Way: An Autobiography 1915–1965*. Jonathan Cape.
Weart, S. R. (1988). *Nuclear Fear*. Havard University Press.

Wells, H. G. (1914). *The World Set Free*. Macmillan.
Werskey, G. (1978). *The Visible College*. Allen Lane.
Wiener, M. (1981). *English Culture and the Decline of the Industrial Spirit, 1850–1980*. Cambridge.
Wilson, D. (1983). *Rutherford: Simple Genius*. Hodder and Stoughton.
Winch, D. (1969). *Economics and Policy*. Hodder and Stoughton.
Wise, L. (n. d.). *Frederick Soddy*. Great Money Reformers No. 3. Holborn Publishing and Distributing Company.

Articles (including printed primary sources)

Anon. (1911). The Glasgow and West of Scotland Technical College. In *Stothers Glasgow, Lanarkshire and Renfrewshire Annual*, pp. 25–27.
Anon. (1914). Report of the inaugural lecture at the University of Aberdeen. *Aberdeen Daily Journal*, 16 October 1914, p. 3.
Anon. (1931). *Journal of Chemical Education*, 8, 1245–6.
Anon. (1904). Evolution of the Atom. *The Morning Herald* (Perth), 25 July 1904, p. 6.
Asimov, A. (1964). *Biographical Encyclopaedia of Science and Technology*, Entry no. 398. New York.
Badash, L. (1979). The Suicidal Success of Radiochemistry. *The British Journal for the History of Science*, 12, 3, No. 42, 245–56.
Beaver, S. H. (1962). The Le Play Society and Field Work. *Geography*, 47, 225–40.
Burchfield, J. D. (1980). Kelvin and the Age of the Earth. In *Darwin to Einstein: Historical Studies on Science and Belief*, (ed. C. Chant and J. Fanvel). Longman, Harlow pp. 180–94.
Clark, C. (1956). The Aberration of Genius. The Work of Frederick Soddy. *The Tablet*, 13 October 1956, pp. 298–200.
Cruikshank, A. (1979). Soddy at Oxford. *The British Journal for the History of Science*, 12, 3, 277–88.
Davies, M. (1992). Frederick Soddy: The Scientist as Prophet. *Annals of Science*, 49, 351–67.
Edgerton, D. E. H. Science and War. *Companion to the History of Modern Science*, (ed. R. C. Colby *et al.*), pp. 934–45. Routledge.
Emsley, J. (1986). Review of Frederick Soddy (1877–1956). *The New Scientist*, April 1986, 62.
Farber, E. (1963). *Nobel Prize Winners in Chemistry 1901–1961*, pp. 400–1.
Fleck, A. (1956). Obituary of Frederick Soddy, FRS. *Nature*, 178, 893.
Fleck, A. Frederick Soddy. *Dictionary of National Biography*, Supplement 1951–60.
Fleck, A. Frederick Soddy. *Biographical Memoirs of Fellows of the Royal Society*, 1957, 3, 203–16.
Freedman, M. I. (1979). Frederick Soddy and the Practical Significance of

Radioactive Matter. *The British Journal for the History of Science*, **12**, 3, No. 42, 257–60.

Hartley, H. (1965). The Contribution of the College Labs. *Chemistry in Britain*, **1**, 521–4.

Jenkins, J. G. (1985). Frederick Soddy's 1904 Visit to Australia and the Subsequent Soddy–Bragg Correspondence: Isolation from Without and Within. *Historical Records of Australian Science*, **6**, 153–69.

Kent, A. (1963). Frederick Soddy. *Proceedings of the Chemical Society*, 327–30.

Lucas, R. (1908). *Bibiliographie der Radioaktiven Stoffe*, pp. 72–3. Leipzig.

Macleod, R. M. and Andrews, E. K. (1971). Scientific Advice in the War at Sea 1915–1917: The Board of Invention and Research. *Journal of Contemporary History*, **6**, No. 2, 3–40.

Page, K. R. (1979). Frederick Soddy: the Aberdeen Interlude. *Aberdeen University Review*, **162**, 127–48.

Paneth, F. A. (1950). Review of Frederick Soddy's *The Story of Atomic Energy*. *Nature*, **1665**, 799–800.

Paneth, F. A. (1957). A Tribute to Frederick Soddy. *Nature*, **180**, 1085–7.

Pattison, M. (1983). Scientists, Inventors and the Military in Britain, 1914–1919: The Munitions Inventions Department. *Social Studies of Science*, **13**, 521–68.

Raynor-Canham, M. F. and Raynor-Canham, G. W. (1990). Pioneer women in nuclear science. *American Journal of Physics*, **58**, 1038.

Russell, A. S. (1956). F. Soddy, Interpreter of Atomic Structure. *Science*, **124**, 1069–70.

Russell, A. S. (1956). *Chemistry and Industry*, No. 47, 1420–1.

Russell, J. (1960). The Le Play Society. The Presidential Address at the concluding Conference of the Society, 9 to 13 April 1960, published as a Report by the Trustees of the Frederick Soddy Trust, p. 5.

Selove, R. E. (1989). From Alchemy to Atomic War: Frederick Soddy's 'Technological Assessment' of Atomic Energy, 1900–1915. *Science, Technology and Human Values*, **14** (2), 163–94.

Shillan, D. (1980). Geography Benefits from a Chemist. *Geographical Magazine*, November 1980, pp. 99–101.

Simmons, W. (1946). H. G. Wells as Atomic Seer. *Morning Herald* (Sydney, Australia), 23 February 1946, p. 6.

Soddy, F. (1966). The Origins of the Concept of Isotopes (Nobel Lecture, 12 December 1922). In *Nobel Lectures including presentation speeches and laureate's biographies: Chemistry 1901–1921*, pp. 370–400. Amsterdam.

Stephenson, H. H. (1914). *Who's Who in Science*, p. 535.

Strachey, J. (1928). Soddy Goes Socialist. *Socialist Review*, July 1928, pp. 1–8.

Trenn, T. J. (1975). Frederick Soddy. In *Dictionary of Scientific Biography*, (ed. C. G. Gillispie), pp. 504–9. Charles Scribner's Sons, New York.

Trenn, T. J. (1979). The Central Role of Energy in Soddy's Holistic and Critical Approach to Nuclear Science, Economics, and Social Resposibilty. *The British Journal for the History of Science*, **12**, 3, No. 42, 261–276.

Turner, F. M. (1980). Public Science in Britain, 1880–1919. *Isis*, **71**, 589–608.

Varcoe, I. (1970). Scientists, Government and Organised Research in Great Britain 1914–1918: The Early History of the D.S.I.R. *Minerva*, **8**, 192–216.

Weart, S. R. (1985). The Heyday of Myth and Cliche. *Bulletin of Atomic Scientists*, **41** (7), 38–43.
Werskey, P. (1971). British Scientists and 'Outsider' Politics, 1931–1945. *Science Studies*, **1**, 67–83.
Wheeler, P. T. (1967). The Development and Role of the Geographical Field Group. *The East Midland Geographer*, **4**, 3, No. 27, 185–95.
Williams, T. I. (1964). Frederick Soddy and the Concept of Isotopes. *Endeavour*, **23**, 54.
Williams, T. I. (1969). Frederick Soddy. *Biographical Dictionary of Scientists*, 482.

Index

Aberystwyth, University College of Wales at 15–17, 30
Adler, Arthur 136
Admiralty Board of Invention and Research 72, 73
alchemy 30
Alembic Club 90
alkali boiler 71
Association of Scientific Workers 87
atomic bomb 162
Australia 38

Beilby family 53
  Emma 57
  George [later Sir George] 45, 46, 54, 73
    death of 127
  Winifred [later Mrs Soddy] 57, 89
    death of 103, 156
    marriage 59
Board of Trustees of the Frederick Soddy Trust 197
Bohr, Niels 90
Borodovsky, V. A. 49
Bragg, William [later Sir William] 59, 73, 86, 87
Branford, Victor 186
Brighton 169
British Science Guild 76, 77, 98
Bunker, Peter 191, 197

Cambridge 'claque' 183
Cameron, Irving 24
Carnegie Trust for the Universities of Scotland 77
Carpenter, Harold Cort 14, 17–18, 20, 45, 60, 72, 73, 90
  death of 157
Cavendish Laboratory, Cambridge 19, 85
Chandos Group 137, 138, 140, 148
Christ Church College 91
Clarendon Laboratory, Oxford 19, 94, 96
Cole, G. D. H. 80, 93, 114, 115, 137, 142
Collie, Dr C. H. 89
Communist Party of Great Britain 87
Cooper, Valerie 136, 147
Cranston, A. J. 46, 65, 71
Curie, Marie 51, 90

Department of Scientific and Industrial Research 73, 86, 95
Ditchling [Sussex] 134, 152, 172
Douglas, Major C. H. 115, 116, 120, 135, 140
Dr Lee's Professor of Chemistry 84, 88
Dyson Perrins Laboratory, Oxford 94

Eastbourne 11, 34, 181
Economic Reform Club and Institute 159, 169
Einstein, Albert 164, 180
Eleventh Hour Flying Clubs 141
Elkins, Prof. T. H. 200

Fajans, Kasimir 49, 50
Farnell, Lewis R. 91
fascism 122–3, 139–40
Fisher, Admiral Lord 73
Fleck, Alexander 47, 49, 90
Frederick Soddy Research Fellowship 199–201
Frederick Soddy Trust, school expeditions 202
Fuller, Major-General J. F. C. 152, 148, 170

Gardiner, Rolf 122, 140
Geddes, Patrick 109–11, 134, 141, 186
Gesell, Silvio 119, 120
Gorky, Maxim 101
Gray, F. W. 63
Gregory, Sir Richard 76, 87, 116, 162, 168

Haber, Fritz 2, 67
Hahn, Otto 51, 185
Haldane, J. B. S. 70, 87
Hall, David 202
Harrington, Prof. Bernard James 25
Hartley, Harold 89, 90, 93
Hevesy, George 49, 185
Hinshelwood, Harold 158
Hiroshima 162
Hitchins, Miss A. F. R. 46, 65, 70, 71
Hobson, S. G. 114

INDEX

Hogben, Lancelot 87
Holloway, Edward 159, 160, 169
Hooke's Joints 103, 156, 191
Howorth, Muriel 20, 178, 179, 80, 101, 183, 184, 185, 191, 193
Hughes, R. E. 14
Huxley, Julian 87

ILP [Independent Labour Party] 53, 78
Institute of Atomic Information for the Layman 180, 182
isotopes 49–50, 185

Japan 101
Japp, Prof. 65
Jex-Blake, Sophia 49, 57

Kelvin, Lord 51–2
Kibbo Kift Kin 122–3
Kitson, Arthur 115, 116, 119, 142, 159
Knox, J. 65

Labour Party 72, 79, 85, 95
Lake District 18
Le Play Society 115, 178, 186, 198
Le Play Tours 187
Levy, Hyman 87, 98
Lewis, Margaret 190, 197
Lindemann, F. A. [later Lord Cherwell] 90, 92, 93, 95, 96
Llewelyn Davies, Margaret 101
London Board to Promote the Extension of University Teaching 38

MacDonald Laboratories at McGill University 25
McGill University 25, 26–30, 31–5
MacKenzie, T. D. 46
Mainau Statement 185
Mairet, Philip 134, 135, 137, 148, 149
Martinez-Alier, Juan 109
Mead, Prof. W. R. 202
Merton College, Oxford 15
Mitrinovic, Dimitri 134–8, 140–1, 144–5, 147–52
  death of 172
monazite sands 51, 100, 158–9
Moseley, H. G. J. 70
Mosley, Oswald 123, 140
Mumford, Lewis 110, 111, 137, 186
Munitions Inventions Department 73
mustard gas production 74–6

Nagasaki 162
Naples 20
National Guilds League 87, 113–15
National Union of Scientific Workers 72, 85–7
New Age 115, 133
New Atlantis Foundation 152
New Atlantis Group 147, 169, 172
New Britain Group 108
  Leamington Spa Conference and split 150–1
  programme 146–7
New Britain Quarterly 144–6
New Europe Group [NEG] 108, 134, 143, 166, 171
  lectures and speakers 143–4
New Zealand 40
Nobel Laureates 184
Nobel Prize 90, 100
novocain 71

Old Chemistry Department, Oxford 94, 100
Oppenheimer, J. Robert 163
Orage, A. R. 114–16, 121–2, 133–6, 140
Ottawa 26
Ouroboros 31, 56
Oxford Union 18
Oxford University Junior Scientific Club 18

radio-chemistry 85, 88
radium 34, 50, 90, 100
Ramsay, Sir William 19, 34, 39, 46, 60
  work with Soddy 34–8
Reckitt, Maurice 137
Rectorial Candidate for Aberdeen University 79
Renaissance Club 169–70
Royal Geographical Society 202
Royal Society 87, 88, 102–3
Ruskin, John 112, 118
Russell, Bertrand 49, 166
Russia 37, 57
Rutherford, Ernest 18, 47, 60, 67, 69, 70, 73, 85, 90, 102
  collaboration with Soddy 31–3, 182–4
  death of 157
  debate with Soddy 1901 26–9

San Francisco 41
science, teaching of 19–20
Shillan, David 189, 197, 202
Sidgwick, Nevil 89, 90, 93
Snow, C. P. 73
Social Credit 108, 116, 119, 121, 138

national conference 1925 121–2
Sociological Society 108, 112, 115, 186, 187
Soddy, Frederick
  Aberdeen inaugural lecture 66
  Aberystwyth period 15–17
  and anti-semitism 183–4
  Australian visit 41
  death of 191
  death of father 41
  and ecology 7–8, 54–7, 67–9, 138
  election as Fellow of the Royal Society 60
  and energy economics 113, 121
  family 11–13
  and Le Play Society 189–91
  marriage 59
  and mathematics 156–7
  Merton period 17–21
  and New Atlantis 173
  obituaries 5–7
  and politics 53–4, 59
  and post-war atomic developments 166–7
  Ramsay, work with 34–8
  and religion 12, 53–4
  research students 46
  retirement 158
  Rutherford, collaboration with 31–3, 182–4
  schooling 13–16
  and Second World War 161–2
  and social responsibility of science 77, 85–7
  on socialism 128
  and Special Contribution court case 168–9
  and university politics 48, 77, 96–8, 92–3, 99–100
  Will of 191
Soddy, Marjorie 197, 202
Soden, D. G. 164
Soden, Helen 134, 148
Stewart, A. W. 103
Strachey, John 128
Sussex, University of 199
Symons, W. Travers 137
Szilard, Leo 68, 164

Tatton, Margaret 178, 186–8, 197, 202
Thomson, J. J. 27, 35, 90
Todd, Dr Margaret 49, 57
Toronto 24, 25
Toronto University 21
transmutation of the elements 55–7
Travancore 51, 158–9

U.S.S.R. 101, 102
undergraduate teaching, Oxford 92
undergraduate teaching, Scotland 46, 66

virtual wealth 124, 125

Wallis, Dr Helen 202
Watts, Alan 149
Wells, H. G. 13, 56, 67–9, 72, 77, 80, 87, 90, 101, 162, 164, 171, 186
women's suffrage 57–9
Wrugh, Reginald 168

way he chose to work, and partly in the character of the man himself.'
He remained outside all institutional frameworks, never laying claim
to 'any particular body of knowledge'. At the same time 'he was a
restless "entrepreneur" in the newly-developing social sciences, who
preferred to test his ideas in personal debate which tended to give
him the advantage.'[6] Although in the area on which he spent most
of his energies, town planning, he remains in the shadow of his better
known follower, Lewis Mumford, Geddes's ideas have gained some
recognition over the last decade or two, especially in the field of the
history of ecological writings. 'He saw the most fundamental question
challenging the present and future generations as the relationship of
man with his natural environment . . . Knowledge in the natural and
physical sciences had the potential to change completely the traditional
equilibrium between human society and the environment.'[7]

This sounds remarkably like Soddy's basic attitude to his subject,
but there were closer similarities as well. They were both to become
'outsiders', geographically and professionally. They had both spent some
years in Scotland, Geddes teaching for part of the year in Dundee and
Soddy as Professor in Aberdeen; they both developed social theories;
they each produced plans for reconstruction and their paths were
to cross again in New Britain. Their plans for reconstruction took
different directions. As described above, Soddy's concentrated on the
contribution of science to the new, socialist Britain, while Geddes's was
wider ranging. Within these wider plans, however, Geddes emphasised
two particular objectives which relate to Soddy's development. One of
these was economic reform, and will be considered below. The other
pointed to the importance of education and the role of young people.
After 'crushing the Prussians', rebuilding the war-torn areas of France
and Belgium and the 'industry-torn areas of Liverpool, Chicago and a
hundred other cities' Geddes wanted[8]

. . . to free schools and universities everywhere from their evil burden of
lifelessness, of Germanic overspecialisations and repressive systems of 'cram and
examine'. Upon the new 'University Militant' would fall the chief responsibility
for guiding world-construction . . .

In 1916 Geddes had spoken to the Trustees of the New Hindu University
at Benares on *University Ideals* and had suggested that[9]

. . . as the universities awaken to their resources and opportunities they will
increasingly co-operate with their cities and regions in order to equip the youth
of each generation more fully and nobly, not only for the struggle, but also for
the culture, of life, and this at all levels . . .

This faith in the powers of education and youth was echoed in a letter
from Soddy to Geddes in which he bemoaned his lack of success in

this period have been dismissed by the 'enormous condescension of posterity' as cranks and failures, but to contemporaries without the benefit of hindsight, any one of these leaders might have been 'right'. Who could have foreseen in 1920 that Freud not Adler, J. M. Keynes not Major Douglas, and the Bloomsbury circle around the Woolfs and Bells, not that around Mitrinovic and the Ditchling Group, would be regarded as successful by historians?

The failure of Soddy's economic and monetary theories to gain acceptance in the short term and, more importantly, to be 'tested' in real situations presents us with difficulties. Marginalised by the 'already strongly professionalized' economists, and rejected by the largely technocratic politics of the inter-war period, he was pushed outside the bounds of respectable thought.[4] Although he himself gloried in the name 'monetary crank' it was hardly a title likely to earn him much respect, especially given that his fellow scientists were, as we have seen, increasingly hostile or indifferent to him and his work. This marginalisation has in recent years been partly corrected by a wave of interest in aspects of Soddy's monetary theory by a new generation of economists concerned with the relationship between economic theory and environmentalism. The most important of these is Juan Martinez-Alier.[5] To Martinez-Alier, Soddy presents the fundamental critique of 'growth' economics. This points to one of the fundamental difficulties of Soddy's monetary theories, which is the extent to which they were linked to what would now be described as his environmental concerns. At the root of all his economics is the recognition, drawn from his scientific research, that natural resources are limited and are linked to economic development. Unfortunately, at the time he was writing, such ideas had not been developed to the point where he could utilise them so that he found himself groping towards a moral economic analysis without the necessary basic tools. However, there are other problems. Soddy's economic thought is sometimes difficult to follow. His language is obscure and his references often too wide ranging. His literary style retains many of the Oxford mannerisms of his youth, moving from classics to thermodynamics and back to the Bible. He embarks on debates with figures who were obscure at the time and are now completely forgotten. As a result, it is often hard to get to the core of his theories. In what follows, I see Soddy's theories within the context of political action and those who wish to follow more closely the theoretical aspects of his monetary theories are advised to turn to Martinez-Alier.

One of the earliest and most influential of Soddy's post-war associates was Patrick Geddes, the sociologist and town planner. Although one of the better known of Soddy's acquaintance outside the bounds of natural science, Geddes has until recently remained an almost forgotten 'outsider'. As his recent biographer explains, 'the reasons for this lie partly in the

CHAPTER 6

Monetary and other theories

For many of the inhabitants of England, the armistice gave hopes of a new beginning, the start of a period of peace and prosperity in the 'Homes Fit For Heroes' which would be provided for all those who had fought in the war. Less certain was the precise mechanism through which this would be achieved. Problems of demobilising the servicemen and women, of dismantling munitions and other war time industries and returning them to peacetime production, and of facing the enormous shortage of housing after years of neglect and lack of new building all gave some sense of the problems which lay in store. Despite the boom of the first year of peace, signs of impending disaster were soon growing. The backlog of demand for various goods in short supply during the war was accompanied by high government spending and the lifting of wartime controls and soon demand outstripped production, prices and labour costs rose and after April 1920 the boom developed into a 'speculative ramp'.[1] A result of this was that 'the period between the end of the First World War and the 1931 crisis was one of the most confused, frustrating, and unsuccessful periods in the history of economic policy-making in Britain.'[2] However, not only economic factors contributed to the immediate post-war society. In addition, 'the direct consequences [of the war] . . . reacted with a multiplicity of other forces creating a tremendous range of side effects and indirect consequences . . .'[3]

These pressures, together with the problems at Oxford detailed in the previous chapter, led Soddy into a search for an explanation of the war and its effects which would deliver him eventually into the company of the 'new economists' and the 'monetary cranks' and almost accidentally into enduring friendships which would help him to survive the loneliness of widowhood and his early retirement. Through the period from the end of the war until the beginning of the 1930s his search took him into a brief association with the Sociological Society and then into National Guilds and the Social Credit Movement, together with a number of connected groups, before he and his monetary theories found a relatively permanent home in the New Britain and New Europe Groups. This chapter will concentrate on the early (and less successful) part of his search which took place during the 1920s, in the course of which he formulated and publicised his theories without finding an audience by whom he felt appreciated. Many of the acquaintances and colleagues he made in

Bodleian Library, Oxford, MS Eng Misc b180; The Origin of the Conception of Isotopes. Lecture given at the Royal Institution, 1 May 1923.
79. Professor Frederick Soddy Papers, Bodleian Library, Oxford, MS Eng Misc b176.
80. Professor Frederick Soddy Papers, Bodleian Library, Oxford, MS Eng Misc b187, b188.
81. Tanaka, M. and Yamasaki, K. (1986). Early Studies of Radioactivity and the Reception of Soddy's Ideas in Japan. In *Frederick Soddy (1877–1956): Early Pioneer in Radiochemistry*, (ed. G. B. Kauffman), pp. 141–54. Reidel, Dordrecht.
82. Smith, D. C. (1986). *H.G. Wells: Desperately Mortal*, pp. 270–3. Yale University Press, Newhaven.
83. See the article by Krivomazov in *Frederick Soddy (1877–1956): Early Pioneer in Radiochemistry*, (ed. G. B. Kauffman), p. 132. Reidel, Dordrecht.
84. I am grateful to Ms Sybil Oldfield for this reference.
85. See the article by Krivomazov in *Frederick Soddy (1877–1956): Early Pioneer in Radiochemistry*, (ed. G. B. Kauffman), pp. 134–6. Reidel, Dordrecht.
86. *The Daily Telegraph*, 3 December 1935.
87. This account is drawn from the records of the Royal Society of London, B.2 and B.3; Professor Frederick Soddy Papers, Bodleian Library, Oxford, MS Eng Misc b170 10 ff125–54.
88. See, for example, the note sent by Dr E. J. Bowen to Muriel Howorth, Professor Frederick Soddy Papers, Bodleian Library, Oxford, MS Eng Misc 170 1.
89. Patent No. 13249, 5 May 1938.
90. *The Times*, 17 December 1956, p. 8.
91. Hall, Sir D. et al. (1935). *The Frustration of Science*, pp. 7–9. George Allen and Unwin.
92. Professor Frederick Soddy Papers, Bodleian Library, Oxford, MS Eng Misc b189.

50. Smith, F. W. (The Earl of Birkenhead) (1961). *The Prof in Two Worlds*, Chapters 4 and 5. Collins.
51. Cole, Dame Margaret (1971). *The Life of G.D.H. Cole*, pp. 70–1. Macmillan.
52. Oxford University Archives, MR/7/2/7.
53. Oxford University Archives, MR/7/2/7.
54. Oxford University Archives, MR/7/2/7.
55. Smith, F. W. (The Earl of Birkenhead) (1961). *The Prof in Two Worlds*, Chapter 4, *passim*. Collins.
56. Speech by Brewer, text in Professor Frederick Soddy Papers, Bodleian Library, Oxford, MS Eng Misc b173 f25.
57. Letter to the author from Dr Collie, 6 March 1991. In the author's possession.
58. Oxford University Archives, MR/7/2/7.
59. *Pioneer*, p. 211.
60. Quoted in Smith, F. W. (The Earl of Birkenhead) (1961). *The Prof in Two Worlds*, pp. 85–6. Collins.
61. Soddy, F. (1919). To the new launch. *The Crucible*, May 1919. Reprinted in Soddy, F. (1920). *Science and Life: Aberdeen Addresses*, pp. 177–8. John Murray.
62. Snow, C. P. (1961). *Science and Government*, pp. 18–21. Oxford University Press.
63. Smith, F. W. (The Earl of Birkenhead) (1961). *The Prof in Two Worlds*, p. 91. Collins.
64. Smith, F. W. (The Earl of Birkenhead) (1961). *The Prof in Two Worlds*, pp. 89–93. Collins.
65. *Pioneer*, p. 215; Cruikshank, A. D. (1986). Soddy at Oxford. In *Frederick Soddy (1877–1956): Early Pioneer in Radiochemistry*, (ed. G. B. Kauffman), p. 163. Reidel, Dordrecht.
66. Cruikshank, A. D. (1986). Soddy at Oxford. In *Frederick Soddy (1877–1956): Early Pioneer in Radiochemistry*, (ed. G. B. Kauffman), p. 159. Reidel, Dordrecht.
67. Oxford University Archives, MR/7/2/7.
68. Minute Book of the Sub-Faculty of Chemistry, Museum of the History of Science, Oxford, 5 Dec 1923.
69. Oxford University Archives, MR/7/2/7.
70. Oxford University Archives, MR/7/2/7.
71. Oxford University Archives, MR/7/2/7.
72. Oxford University Archives, MR/7/2/7.
73. *Nature*, 8 December 1923, 839.
74. *Scientific Worker*, December 1924, 86–7.
75. Oxford University Archives, MR/7/2/7.
76. Professor Frederick Soddy Papers, Bodleian Library, Oxford, MS Eng Misc b170 f497.
77. *Pioneer*, p. 215.
78. See, for example, Contribution to discussion on isotopes. (1921). *Proceedings of the Royal Society*, A **99**, 97; Isotopes. Introduction to discussion on Atomic Structure. Reunion de Bruxelles, April 1922; The Nobel Lecture: The Origin of the Concept of Isotopes. (1923). In *Les Prix Nobel en 1921–1922*. Stockholm; Presentation plan and lecture text in Professor Frederick Soddy Papers,

(1986). Soddy at Oxford. In *Frederick Soddy (1877–1956): Early Pioneer in Radiochemistry*, (ed. G. B. Kauffman), p. 159. Reidel, Dordrecht.
26. Davies, M. (1992). The Scientist as Prophet. *Annals of Science*, **49**, 351–67, this reference to 356.
27. Letter from Dr C. H. Collie to the author, 4 February 1991. In the author's possession. Dr Collie was later Lee's reader at Christchurch, succeeding Dr A. A. Russell.
28. Letter from Dr C. H. Collie to the author, 8 August 1991. In the author's possession.
29. Professor Frederick Soddy Papers, Bodleian Library, Oxford, MS Eng Miscx b 171 f39.
30. Personal communication from Margaret Shillan.
31. Membership and Meetings Book of the Alembic Club. In Museum of the History of Science, Oxford MS Museum ISO.
32. *Pioneer*, p. 209.
33. Professor Frederick Soddy Papers, Bodleian Library, Oxford, MS Eng Misc b170 ff344–482, f493.
34. Professor Frederick Soddy Papers, Bodleian Library, Oxford, MS Eng Misc b170 f17, ff344–482, b171 f33.
35. Cruikshank, A. D. (1986). Soddy at Oxford. In *Frederick Soddy (1877–1956): Early Pioneer in Radiochemistry*, (ed. G. B. Kauffman), p. 159. Reidel, Dordrecht.
36. Smith, F. W. (The Earl of Birkenhead) (1961). *The Prof in Two Worlds*, p. 83. Collins.
37. Letter from Dr C. H. Collie to the author dated 8 August 1991. In the author's possession.
38. Letter from Dr C. H. Collie to the author dated 8 August 1991. In the author's possession.
39. Letter from Dr C. H. Collie to the author dated 8 August 1991. In the author's possession.
40. Letter from Dr C. H. Collie to the author dated 8 August 1991. In the author's possession.
41. Letter from Dr C. H. Collie to the author dated 28 July 1991. In the author's possession.
42. Letter from Professor R. P. Bell to the author dated 3 February 1991. In the author's possession.
43. Farnell, L. R. (1934). *An Oxonian Looks Back*, pp. 303–5. Martin Hopkinson Ltd.
44. Smith, F. W. (The Earl of Birkenhead) (1961). *The Prof in Two Worlds*, pp. 81–4. Collins.
45. Oxford University Museum of the History of Science, MSS MUS 135–6.
46. I am particularly grateful to Mr A. V. Simcock in the Old Museum, Oxford University, for drawing my attention to these questions.
47. Speech by Brewer, text in Professor Frederick Soddy Papers, Bodleian Library, Oxford, MS Eng Misc b173 f25.
48. Letter from Dr C. H. Collie to the author dated 28 July 1991. In the author's possession.
49. Cruikshank, A. D. (1986). Soddy at Oxford. In *Frederick Soddy (1877–1956): Early Pioneer in Radiochemistry*, (ed. G. B. Kauffman), pp. 159, 163. Reidel, Dordrecht.

Notes and references

1. Emsley, J. (1986). Frederick Soddy. *New Scientist*, 10 April 1986, p. 62.
2. Soddy, F. (1918). Review of *The Life of Sophia Jex-Blake* by Margaret Todd. *Nature*, 15 August 1918, p. 11.
3. Badash, L. (1986). The Suicidal Success of Radiochemistry. In *Frederick Soddy (1877–1956): Early Pioneer in Radiochemistry*, (ed. G. B. Kauffman), pp. 27–41. Reidel, Dordrecht.
4. Freedman, M. I. (1986). Frederick Soddy and the Practical Significance of Radioactive Matter. In *Frederick Soddy (1877–1956): Early Pioneer in Radiochemistry*, (ed. G. B. Kauffman), pp. 171–6. Reidel, Dordrecht.
5. Rutherford Papers, Cambridge University Library, AD 7653, S176, S177.
6. Reprinted as: The Ideals of a Science School. Soddy, F. (1920). *Science and Life: Aberdeen Addresses*, pp. 181–206. John Murray.
7. Soddy, F. (1920). *Science and Life: Aberdeen Addresses*, p. 187. John Murray.
8. Soddy, F. (1920). *Science and Life: Aberdeen Addresses*, p. 193. John Murray.
9. For a history of the DSIR see Varcoe, I. (1974). *Organising for Science in Britain: A Case Study*. Oxford University Press.
10. Varcoe, I. (1974). *Organising for Science in Britain: A Case Study*, p. 22. Oxford University Press.
11. See, for example, *Nature*, 24 July 1919, p. 404.
12. *New Statesman*, Special Supplements, 1917.
13. For a short history of the NUSW see Werskey, G. (1978). *The Visible College*, pp. 39–41. Allen Lane; Outline of the History and Policy of the National Union of Scientific Workers. *Scientific Worker*, **7**, February 1926; The Association of Scientific Workers, Ten Years Work. *Scientific Worker*, **9**, October 1928.
14. Quoted in Werskey, G. (1978). *The Visible College*, p. 40. Allen Lane.
15. See his testimonials from Cole and others for the rectorial campaign at Aberdeen, described in the previous chapter.
16. Werskey, G. (1978). *The Visible College*. Allen Lane; Rose, H. and Rose, S. (1969). *Science and Society*, Chapter 3. Allen Lane.
17. Letter to *The Observer*, 26 September 1920.
18. Lecture to National Union of Scientific Workers, May 1920. Professor Frederick Soddy Papers, Bodleian Library, Oxford, MS Eng Misc b172.
19. *Nature*, 6 May 1920, pp. 309–10.
20. *Scientific Worker*, December 1920, **1** (6).
21. Royal Society Council Documents, CD 1230. Other documents in this case are contained in CD 1208–32.
22. See Morrell, J. (1994). The Non-medical Sciences. In *The History of the University of Oxford. Vol. VIII: Twentieth Century Oxford*, (ed. B. Harrison). Oxford University Press; Green, V. H. H. (1974). *A History of Oxford University*, Chapter 11. Batsford; Smith, F. W. (The Earl of Birkenhead) (1961). *The Prof in Two Worlds, passim*. Collins.
23. Oxford University Archives, MR/7/2/7.
24. *Pioneer*, p. 212.
25. Story told by Dr F. M. Brewer at: The Meeting of Commemoration held on the Eve of the First Anniversary of the Death of Frederick Soddy. Published as a pamphlet by the New Europe Group. Also told in Cruikshank, A. D.

the Council, which took refuge in the procedures of any official body, asking for legal opinion, and referring the questions to the Archbishop of Canterbury, Viscount Hailsham, Viscount Halifax and Sir John Simon who all agreed that any change such as those proposed would require a change to the Charters of the Society which had been drawn up in 1662 and 1663. Against the opinion of such august personages, it is not surprising that Soddy's candidates lost the election, and the question was allowed to disappear.[87]

Soddy's other area of interest during this period, mathematics, is also in some ways a culmination of many of his earlier interests. Throughout his scientific career, he insisted on the importance of the 'physical', by which he meant the tangible. He was also an extremely able experimental scientist, showing a high level of manual dexterity and an ability to construct experimental apparatus.[88] However, he was also able to be attracted by a particular problem and to work on its solution to the exclusion of anything else. This seems to have happened in 1935 when he attended an 'Exhibition to Celebrate Hooke's tercentenary' at Oxford's Museum of Science. Three years later he patented his 'Improvements in Hooke's Joints and Universal Couplings'.[89] His interest in this was maintained until the end of his life and he left £1000 in his will to develop its educational and other uses.[90]

Also in 1935 Soddy wrote his own unwitting 'farewell' address to science, when he contributed the 'Foreword' to the influential collection of essays *The Frustration of Science*.[91] In this he repeats his warnings of the dangers of science. 'Unfortunately, scientific powers of inflicting mass suffering are now so powerful that once started they are hardly likely to stop so long as there is anyone left to suffer.' He ends with an old message, learned again with bitterness at Oxford, that the public 'require its universities and learned societies should . . . do that for which they are supported, in cultured release from routine occupations, and speak the truth though the heavens fall.'

Soddy resigned his professorship at Oxford in 1936 following the death of his wife. Looking back he can have felt little satisfaction. There was almost no research, rather more influence from writing and personal contacts, but above all absolutely no sign of the kind of institute that it had been hoped he would found. A large part of the blame for this must be borne by the authorities and faculty of the university, who were never supportive and often deliberately destructive. A letter of sympathy from his successor at Glasgow, A. W. Stewart, must have been welcome. Stewart wrote that 'it needed no great penetration to guess that the Oxford people have not gone out of their way to make you comfortable there . . . one can imagine they have not been altogether congenial colleagues.'[92]

Finally, in 1924, Soddy, along with Rutherford and a number of other outstanding scientists, was elected a member-correspondent of the USSR Academy of Sciences but his works, especially the early ones, have continued to be studied in the USSR where his contribution to the science of radioactivity has been more consistently recognised than in Britain.[85]

After this brief interval of comparative success, the early 1930s, Soddy's last years at Oxford, provide a dismal story. There is no evidence of any attempt at scientific work, even of the limited kind he continued through the twenties, although there was a new edition of *The Interpretation of the Atom*, published in 1932. Instead, as will be discussed in a later chapter, his interests outside Oxford took more and more of his energy, while within what may be loosely defined as 'science', two different areas came to dominate his thoughts. Firstly, he returned to his concerns with the professional organisation by entering into a dispute with the Royal Society, and secondly he developed a fascination with the solution of particular mathematical problems.

The dispute with the Royal Society was seized upon by the newspapers, and the publicity led to a suggestion that it was more important than the facts actually show. In 1935 Soddy canvassed support for a move he proposed regarding the regulations governing the election of new members to the Council of the Royal Society. He said that he believed that the Council had become effectively self-perpetuating, the present members electing their own successors so that no ordinary Fellows could ever reach the Council. The result was, he said, that[86]

under present conditions the Royal Society is not a national society representing science to the nation. There is no democratic control, and Fellows are not consulted. A small group has a stranglehold on the society and seems to think it rules by divine right. Anyone with the courage to oppose it is a marked man. The group can almost destroy him.

Compared with the violence of the complaints, the reforms demanded seem rather mild. They included a reform of the voting system, and a concentration on matters of science as a whole, including the relationship between scientific discovery and the community. Of these, the reform of the system of electing members of Council was the controversial item, and Soddy canvassed support for his motion, giving lunches, circulating his proposals and finally gaining the support of 91 of the 459 members of the society. He then announced that the election would be contested and produced an alternative list of fourteen names as a possible council. As in so many of his dealings with official bodies during this period, Soddy had considerable and respected support for his position. However, by his belligerent method of attempting reform he seems to have antagonised

The least likely of these was Japan. Satoyasu Iimori, who was later to become 'the founder of radiochemistry in Japan', studied with Soddy during 1920 and 1921, working on the determination of uranium in Ceylon monazite. He also took back to Japan ideas and materials for teaching and setting up a research laboratory. On his return he instigated a course of lectures called 'Chemistry of the Radioelements' after Soddy's book and used instruments and standardised radioactive preparations, which he had brought back from Soddy, to develop systematic studies of radiochemistry in the laboratory of the Institute for Physical and Chemical Research. After this, he turned to interests more like those of Soddy and spent the next twenty years, until the end of the Second World War, in surveying uranium- and thorium-containing minerals on the mainland of Japan and Korea.[81]

The other country was the USSR, where Soddy's influence was possibly greater but less personal. As was mentioned previously, Soddy's works had reached a wide and enthusiastic audience from 1903 onwards, and during the 1920s this was widened further. H. G. Wells, who, as already discussed, had made careful and influential use of Soddy's works, had developed a strong interest in Russia from 1914 when he had visited St Petersburg and Moscow. While there he had met Lenin, who was so impressed that he embarked on reading *The War in the Air* and *The World Set Free*. Wells also renewed his acquaintance with Maxim Gorky. Wells became an important figure in interpreting Russia to the West and in sending scientific books to Russia through the British Association.[82] Through this meeting, Wells introduced Soddy's works to Gorky. As Gorky wrote to Wells, 'All the books have already been reviewed by various professors ... some books, and first of all, the papers and speeches of Soddy, are being translated into Russian.' In addition, in 1923 Gorky hoped to commission papers from Soddy and Wells for the new scientific-humanist journal *Beseda*, but these never seem to have materialised.[83] Also in the 1920s Soddy and Wells were both among the many vice-presidents of The Society for Cultural Relations Between the Peoples of the British Commonwealth and the Union of Socialist Soviet Republics, (sic) and his name still appears in this context in 1925. In 1922 Soddy lectured for the Society and his visit was described by Margaret Llewelyn Davies:[84]

Last Sunday, we had Professor Soddy up in Hampstead and a crowded meeting. Such a nice Scientific person—so simple and intellectual—and unlike the ordinary 'speaker'. I am not sure he is wise to have gone off into Finance. He feels no one has seen the real thing, that causes all the ills, and that if only we got rid of private Banking all would be well. He is so sincere, and can't stand the idiocies of our civilisation.

the humanities and other non-research departments of the university, he now believed that the benefaction of George Aldrich, MD, had been misused. It was, he insisted, intended to provide chemical apparatus and materials or to pay the salary of the demonstrator, whichever was to the greater benefit of the chemical department. Soddy now suggested that the money provided was insufficient for the salary and so could provide only a part of the sum required. This sum could be met by the university, leaving the bequest for other uses, specifically for apparatus and materials which would fulfil the greatest need in the department. With the Aldrich bequest, '. . . it would be possible to procure instruments of precision, and the rarer and more costly apparatus and materials, as articles of platinum and radioactive preparations, which the Grant to the Department from the general funds of the University is insufficient to provide.'[76] As in the Carnegie case, this was a Quixotic tilting at windmills and the case was dismissed, but this time, unlike his experience in Aberdeen, without the support of the British Science Guild or the publicity of even *The Scientific Worker*.

However, in 1928 there was finally some sign that Soddy's fortunes might have changed. A new secretary was appointed to the Curators of the University Chest who was friendly towards Soddy. Perhaps through his influence, or perhaps as reparation for his problems with the Faculty Board, Soddy was awarded £13 000 for the reconditioning of the Old Chemistry Department. He undertook this work himself, accepting no help from architects or other specialists. His design was so well conceived that as late as 1957 a new laboratory could be added on, as he had initially suggested. It is, however, not surprising that it 'absorbed my entire energy for some years'.[77]

Although he now had his laboratory, it was by this time too late for Soddy to regain his position at the forefront of science. His book *The Interpretation of Radium* was brought out in a fourth edition in 1920, but other than articles and contributions to discussions on isotopes at the time of the Nobel Prize[78] he published no other scientific materials at this time. However, despite all the other distractions of his first decade at Oxford he had not entirely abandoned research. Although he had embarked on no new projects, he had continued his search for either a cheaper method of extracting radium from the ore or the discovery of an alternative to radium. During late 1920 and early 1921 his notebooks show that he continued the work on monazite sands which he had begun in Glasgow before the First World War,[79] and from 1922 to 1927 he worked on radium samples from Joachimstal mines.[80] Neither of these attempts was successful.

While his contribution to scientific research during this period was minimal, Soddy's influence was important in two very different countries.

and away from socialism in other respects, which is discussed in the next chapter, in terms of science he remained firmly on the 'left'. His message was still very much that of 1919. Science can bring abundance if government and tradition do not stand in its way. It was a message which must have seemed particularly apposite in view of his continued battles with the Oxford establishment. This aspect of his beliefs never changed. Although his politics moved away from conventional socialisms, his support for the Communist scientists who wrote in *The Frustration of Science* in 1935, and his ambiguities on atomic power during the 1950s, show he never gave up his belief that science in the right hands could liberate humankind.

If these feelings needed reinforcing, in 1927 Soddy's problems with the university reappeared. This time the difficulty was over the changes in regulations of the Aldrichian Praelector in Chemistry, or in simpler terms the problem of the demonstrator had appeared again. The results of these changes were that the demonstrator in Soddy's laboratory could be appointed on the authority of the Faculty Board of the Physical Sciences but without specifically mentioning any rights of the head of the department. Yet again, Soddy's non-membership of this board was causing difficulties. The Vice-Chancellor again suggested a compromise in the form of a change to the relevant statute ensuring that approval of the head of department would be required, but Soddy still objected that this would not be retrospective. It is unclear whether Soddy's objection was entirely one of principle since the candidate for the post was Lambert again, but in any case he continued his case and brought a petition in the Privy Council. Again, he was guilty of misjudgement, however understandable his feelings might have been. Until he actually brought his case, there was considerable sympathy for it. A letter to the Vice-Chancellor from a colleague summed it up: 'it does seem to me that although all parties acted within their rights, it is unfortunate that anyone should be appointed a Demonstrator under any Professor without the Professor being consulted in the matter. It seems to me that Soddy has a real grievance in this instance, but one against the Board of the Faculty of Physical Science of which he is not a member, and which did not consult him . . . It is unfortunate that the head of so important a Department should not be a member of the Board; and there, I fancy comes in the old struggle between the tutors and the Professors.'[75]

What was not made clear in any of his letters was that Soddy by now had a further cause for complaint which was revealed in his actual petition. His reading of the Statutes to check the conditions of the appointment of the Praelector had convinced him that there was another, more serious charge to be answered. Just as, at Aberdeen, he had been convinced that the Carnegie moneys were misappropriated by were loyal to the university and other members of faculty rather than to him as the head of their laboratory. It also was the result of Soddy's frustration at not achieving any of his goals at Oxford. While relations with Lambert and Appleby were temporarily mended, the larger question was still a problem.

The frustration caused by the machinations of the committees, and his clear feeling of impotence against the workings of a system from which he had been deliberately excluded, finally provoked Soddy into taking legal advice about his position. On 3 January 1924, he wrote to the Vice-Chancellor explaining that he believed that the Dr Lee's Professor should lecture on inorganic and physical chemistry, that he had accepted the post on the understanding that his laboratory should be modernised, re-equipped for research and teaching and provided with adequate staff, that under the new scheme there was no provision for practical instruction in physical chemistry, and that the Dr Lee's Professor would be confined to teaching inorganic chemistry, as physical chemistry would be taught in the college laboratories.[71] In response, the Vice-Chancellor hoped that he would be able to soothe Soddy's feelings and come to some kind of agreement. Indeed, such and agreement may have been reached, but it was only temporary.[72]

It was not only that Soddy's problems in relation to Oxford remained unsolved. The debates about the relationships between science, scientists, government and industry which had been a feature of the early 1920s had not been resolved and indeed were worsening as the slump began to make its full effects felt. In 1924 they were brought to public, or at least scientists', notice again by a conference organised by the British Science Guild on 'Science and Labour'. The conference was a meeting of major figures from the world of science and Labour politics, brought together to discuss possible solutions to the increasing problems of unemployment and economic decline. Amongst others, Sidney Webb spoke on the situation of workers and Oliver Lodge explained the benefits to be expected from the application of science, especially radioactivity. However, as Hyman Levy pointed out in his review of the conference for *Nature*, while 'the belief was constantly reiterated that the industrial problem must yield to scientific solution' and there was a great deal of discussion about increasing production, there was little consideration of the role of industrial workers. Levy argued that they might not welcome scientific change which led to unemployment but would look rather to some redistribution of work and wealth.[73]

Surprisingly, Soddy seems not to have attended this conference, and there is certainly no mention of him contributing in any way. However, he reviewed the published account of the proceedings in *Scientific Worker*.[74] This review suggests that, despite his movement towards social credit

The result was that, feeling snubbed by his exclusion from the main committee, Soddy declined the invitation to join. Instead, after receiving the draft proposals of the subcommittee, with which he disagreed, Soddy submitted a far more radical proposal. This was in line with the pre-war suggestions and, in fact, very like those adopted in the late 1930s after his retirement.[66] He recommended that all chemistry teaching should be undertaken by the university and not by the colleges. This would result in advantages for students and especially for the postgraduate, for whom 'the organised resources of the whole university should be at his disposal and not a scattered collection of teachers and laboratories interconnected respectively by a tangled network of personal relationships and private ownerships.' He believed that his proposal could be financed by contributions from all the colleges since it would replace college tutors. When the changes had been organised, benefactors could be sought.[67] Unremarkably, this challenge to the whole system of teaching at the university and to the jobs of all college tutors was rejected by the committee which consisted almost entirely of those whose posts would be at risk. By December 1923 the suggestion put to this subcommittee was that physical chemistry should be taught in the Balliol and Trinity laboratories, organic chemistry in the Dyson Perrins and Queen's College laboratories, leaving inorganic chemistry in the Old Chemistry Department and Christ Church Laboratories.[68] The result would be that Soddy would be forced to teach inorganic chemistry, in which he had little interest, and be excluded from his own speciality, physical chemistry. In all these discussions Soddy's was the only dissenting voice.

At the same time, in December 1923, what should have been a question of relatively unimportant departmental administration was inflated by its connection with the workings of this committee. On his arrival at Oxford, Soddy had inherited a laboratory staff which included three demonstrators, of whom B. Lambert was still employed as the senior in 1923. Lambert and Appleby, who had been appointed as a junior, had both been invited to serve on, or at least give evidence to, the committee. Unfortunately, Soddy, having refused to serve on the chemical subcommittee, considered it a sign of disloyalty that anyone from his laboratory should be involved with it. He went so far as to ask his demonstrators 'to consider the matter further, and to resign either from the Committee or from your Demonstratorship in this Department.'[69] This threat was misjudged. Lambert and Appleby responded to Soddy's letter by involving the Vice-Chancellor who managed to achieve a temporary peace by sending Lambert to Soddy where 'I [Lambert] had a long and amicable discussion with him about the whole situation and I have, for the first time, a clear idea of his point of view'.[70] However, the threat had been provoked immediately by Soddy's very real feeling that his staff his support was able to increase financial provision from the university. Within the next two decades the number of technical and research staff and undergraduate and postgraduate students in the Clarendon Laboratory increased overall from nine to fifty. In addition Lindemann tapped rich outside sources, especially industrialists whom he knew personally or believed to be sympathetic to his plans. His fund-raising efforts culminated in his negotiation with Lord Nuffield who, despite his dislike of pure physics, gave £8000 per annum for eight years from 1946 and then a further £20 000. Thus, despite his arguments with college bodies and unpopularity with colleagues, Lindemann was able to build the Clarendon into a research laboratory of worldwide stature.[64]

In contrast to this Soddy was able to achieve very little, and that little very slowly. The reasons for this are complicated, but centre on his clearly established dislike of bureaucratic and authoritative administrative bodies. As well, although he could exert considerable charm in small, personal gatherings he had no diplomatic or persuasive skills in larger meetings. Anyway, he felt them unnecessary. His belief in himself as a scientist and in his politics led to his enduring conviction that facts and arguments would present a self-evident case without the need for persuasion and discussion. This, together with an almost suicidal disregard for his own position, contributed to his reputation for arrogance. These somewhat unattractive characteristics were all demonstrated within his first ten years at Oxford. Although he had hoped that the reorganisation of Oxford science would be established by a Government Commission of Enquiry, he began, almost on arrival, to make detailed plans for the improvement of his laboratories, but without indulging in the kind of political canvassing which was to make Lindemann's attempts successful. When approached, the university authorities refused funding for Soddy's plans. The reasons for refusal were never made clear. It might have been because they intended to await the findings of the Commission, although a less charitable interpretation would point out that Soddy did not have the kind of influential support that Lindemann found for himself. Whatever the reasons, the refusal actually proved to be merely a delay, since by June 1924, a grant had been made to Soddy of only £100 less than he had requested.[65]

However, this success was only a short interlude and problems of the university were very soon to recur. During 1923 a committee was set up to consider the division of teaching between the colleges and the university. Soddy had not been invited to sit on the committee itself although he was invited to become a member of a subcommittee on 'The Organisation of the Teaching of Chemistry'. This seems a reflection of the absurd situation arising from the clashes between him and Sidgwick which had resulted in the exclusion of Soddy from the Faculty Board.

Under the circumstances, it is not surprising that the staff and students in both departments were few, disheartened and in need of total revitalisation. In all this, Soddy and Lindemann should have been natural allies with a common cause. There were a number of common areas including the shared circumstances of their new posts and shared dispositions which made them personally difficult and generally convinced that their views of any situation were correct. They also shared a belief in the importance of 'pure research'. Lindemann gave this as one of the reasons for wanting the Oxford post; 'the main desideratum, however, is to my mind that there should be ample opportunity for research and not too much time lost in teaching. By research I mean pure research, unfettered by consideration of industrial application . . .'[60] This could have been written by Soddy himself. One of his concerns, especially in the post-war years, was the teaching load on academic staff, especially the juniors, and the extent to which this made research impossible. In his address to the Aberdeen students in May 1919, he had said 'the teaching of a single main science subject, such as chemistry . . . today involves probably more actual work than the teaching given in the whole Faculty of Arts a century ago . . . research, worth the name, is a practical impossibility.'[61] He was also to spend much of the next few years opposing the establishment of the Department of Scientific and Industrial Research because of its emphasis on the application of scientific research to industry. Although there were these similarities, there were also enormous differences. Lindemann's social life was increasingly concentrated outside the university and scientific circles while Soddy's consisted almost entirely of family, colleagues, and friends who were in some way associated with science. Lindemann was something of an athlete, playing tennis to championship level in his youth while Soddy preferred non-competitive climbing and walking. Above all, their politics were diametrically opposed. As we have seen, Soddy was fairly far to the left of the Labour Party, while Lindemann was equally far to the right of the Conservative Party.

The ways in which the new professors began to deal with their problems is a fascinating object lesson in different approaches. Lindemann was not, despite his insistence on pure science, primarily a research scientist. As C. P. Snow perceptively remarked, he knew he was 'not going, by high standards, to make a success of pure science' by 1919. Accordingly, he set out to create a career in high-level scientific administration.[62] Lindemann apparently had all the necessary attributes for such a career, including a desire to move in high society, and 'charm and persuasiveness which he could command when he wished'.[63] He soon began to put these to work. He made an ally of I. O. Griffith, fellow of Brasenose, member of the Hebdomanal Council and most relevant committees, and with

It was, in any case, clear that science in the university was to be reorganised in the relatively near future. In 1909 a questionnaire about the organisation of Oxford science had been sent to members of the scientific professorate in other institutions and the answers had unanimously recommended the reorganisation of science at Oxford into one body or institute with a Director and then professors, junior faculty, and research students as well as undergraduates. The advantages of this were seen to be improved contact between the different branches of chemistry and a more coherent teaching programme.[52] Once accepted, this led to a consideration amongst the various university departments of suitable space for a new laboratory, and in turn to an agreement by the Governing Body of Christ Church to advance the Dr Lee's Readership to a Professorship so long as the university provided laboratory space.[53] Thus the appointment of Soddy and of Lindemann can be seen as part of a long-term reform of Oxford science.

To accommodate the reforms, the construction of new laboratories had begun. In 1913 space and money for a chemical laboratory, the Dyson Perrins, had been found. William Henry Perkin, the new Waynflete Professor (Soddy had unsuccessfully applied for the post), had been consulted over the detail of the plans and had taken almost complete occupancy of the new building.[54] However, the new Dr Lee's professors were less fortunate. Lindemann found that the Clarendon Laboratory was in much the same state as when it had been built in 1872 from the proceeds of the sale of the Earl of Clarendon's *History of the Great Rebellion*. It had an inadequate water supply, lighting by gas burners and no electricity. With these resources Lindemann was expected to build a new research department which would rival the Cavendish.[55] Similarly, Soddy was given some space in the Dyson Perrins for his prelims teaching but otherwise found himself with the 'old, ill-kept, ill-furnished and, so far as research was concerned, completely unequipped laboratory, known rather appropriately, as the Old Chemistry Department'.[56] The staff of these departments was as unsuitable for modern research as the buildings. According to Soddy's demonstrator of the mid-1920s, chemistry was 'a practically derelict department torn by internecine strife'.[57] Although his enduring interest in alchemy might well have been engaged by the part of the laboratory designed by Ruskin and modelled on the Abbot's Kitchen at Glastonbury Abbey, with the latest additions dating from 1875, its condition meant that there can have been no attraction at all in the inheritance of what has been called the oldest and least convenient lecture theatre in Oxford.[58] Soddy's reaction to the place was one of horror. He noticed the appalling 'lack of attention and wholly uncared-for appearance of the place. The flooring cannot have been conditioned for years: the benches were almost rotting and inadequate.'[59]

Without personal meetings (one of the areas where he was usually able to impress acquaintances), Soddy was unable to know or influence his students.[48]

In addition, at least an influential minority of the college tutors had a particular case against Soddy. Sidgwick, although his research ranged widely over a number of fields rather than concentrating on one particular area, was a long-established and well-respected teacher. Hartley, as head of the Balliol–Trinity laboratory, was effectively head of chemistry teaching. Therefore they both had strong expectations of being awarded the Lee's Chair. Hartley's reaction to Soddy's arrival is not recorded, but Sidgwick 'was at little pains to conceal his animosity to the new professor' and since he had a reputation for a bitter tongue and remembering grievances he was almost bound to be involved in disputes with the equally argumentative Soddy.[49]

None of this would have been important if Soddy had found research possible. Oxford dons had a long and zealously guarded reputation for feuds and difficulties in accommodating each other. The experiences of Lindemann, later Lord Cherwell, are remarkably like Soddy's in this respect. On arrival at Wadham in 1919, before both of the Dr Lee's Professors were transferred to Christ Church, Lindemann embarked on a squabble over his rooms which he finally won, thus securing his reputation for obstinacy and the dislike of some of his colleagues. However, he never allowed such relatively minor matters to interfere with his ambitions inside or out of Oxford and so won respect for his success.[50] This is not surprising: G. D. H. Cole's acceptance at Oxford, despite his politics, has been described as due to the way in which 'the old universities preserved a surprising amount of freedom, and even eccentricity for the "gentlemen" admitted to their ranks'. Difficulties arose only for those not so honoured. As a fellow at Magdalen, even Cole's political beliefs could be accommodated once the holder had been accepted as one of the group.[51] However, Soddy never managed to be accepted. He lacked Lindemann's insensitivity to criticism and dislike and Cole's belief in himself and in the Oxbridge system of education, culminating in college fellowships which ensured facilities for research, at least in the humanities where necessary facilities were already in existence. In contrast, Soddy was sensitive to dislike, but reacted truculently; he was politically opposed to the privileges of the system; he was attempting to introduce a new discipline which required expensive new facilities; and, most of all, he was unable to behave with ease and diplomacy with colleagues and bureaucracy. As a result, while the questions of membership of boards and other bodies could have been solved with a little give and take on each side, Soddy refused to move.

the reason, which was never discovered, Farnell and his wife received a box of chocolates which were discovered, on opening, to be 'covered with a repulsive white powder'. Soddy was asked to analyse this powder and found it to be ground glass. Scotland Yard was summoned; the press discovered this story and used it as part of an attack on Farnell; an undergraduate confessed. Soddy wrote to *The Times* with his analysis of the powder and to explain that the culprit had been found. However, this did not satisfy the popular press who suggested that 'Professor Soddy was wrong in his analysis'.[43] This bizarre incident no doubt strengthened the friendship between Farnell and Soddy, but it was unlikely to endear Soddy to the students.

The omission from the faculty board was more serious since this was the forum for discussion about all matters of importance to science and scientists within the university. This was said to be directly connected with Lindemann. Members were required to be MAs of the university. Lindemann was not eligible for this degree since he had not been there as an undergraduate, and was awarded his MA by decree.[44] Soddy, as an undergraduate, was entitled to an MA but was expected to pay for it. He felt this was unfair and, for many years, refused to pay and thus excluded himself from membership. Even after he had paid the necessary small sum his exclusion continued because his admission was vetoed by other members. While apparently trivial, this was to have important consequences since he was never able to be present at meetings where matters of importance to the sciences were discussed, and could never present his own point of view.[45]

Of more importance than this were the problems associated with laboratory space and the appointment of more junior staff. At Oxford in the early 1920s almost all teaching was done by college fellows in college laboratories, with some specialisation between them—the system which had obtained during Soddy's undergraduate years. This was especially true of chemistry which was taught in Balliol and Trinity laboratory (where Soddy had experimented from 1898–99) and in Jesus, Christ Church and Queen's. The university laboratories and faculty were provided in addition to these and were forced to enter an uneven competition for students and resources. Students were initially selected through the colleges by dons who would then teach them. The university faculty gave lectures but there was no opportunity for them to have the close contact with undergraduates to which lecturers in other establishments, and especially Soddy in Scotland, had been accustomed.[46] For Soddy, the result was that he had a relatively heavy teaching load, with responsibility for instruction in both physical and inorganic chemistry, and gave the general course of lectures in physical chemistry for some years, but was rewarded with relatively little contact with the students.[47]

Faculty of Physical Sciences, nor was he eligible to sit on the governing body of Christ Church, of which he was not only a Student, but also a comparatively senior member of the college.[35] The excuse given for the latter constraint was that the governing body of the college was large before the attachment of three new science professors to the college, including the two Dr Lee's—Soddy and Lindemann. That both of these men were to develop reputations for intransigence and overbearing behaviour in discussion cannot, at this stage, have influenced the decision, but both felt excluded from an important part of college life. Lindemann was to remedy this, and over a period of time was accepted, but he lived in college and so was amongst his colleagues all the time.[36] Soddy, in contrast, lived with his wife and seems to have been unable to use this social milieu to remedy his professional and personal exclusion and as a result remained apart.

Although apparently an isolated figure at this time, a rather odd incident shows him to have been accepted at least by the then Vice-Chancellor, Lewis R. Farnell. Soddy was involved with Farnell in two ways. Firstly, he supported Farnell's rather archaic notions of the responsibility of the university for the morals of the students. In 1920 Farnell 'decided to act according to the letter of the law and protect the young men from the dangerous ideas of the time.'[37] Using his power under medieval statute Farnell 'had women of doubtful character removed from Oxford by the Police and insisted that the theatre programme be submitted to him for censorship.'[38] However these were no ordinary undergraduates, 'but young men just returned from the First World War and all imbued in a mild way with a spirit of reform and a determination not to let the "old guard" let them in for another [war].'[39] Ironically, in view of his own feelings about the war, Soddy was dragged into the whole affair on Farnell's side 'out of loyalty' and was pilloried in both the local and national press. 'A Nobel Prize Winner and world figure making a fool of himself was a temptation they could not resist.'[40] It is possible that Soddy's identification with these reactionary causes led to his apparent difficulties with other undergraduate students in the 1920s. He always lectured in 'a formal way in a frock coat . . . this was rightly considered inappropriate for young men just back from the war',[41] and according to an undergraduate student of the period was 'considerably heckled by the audience (not a common thing in those days)'.[42]

Equally unfortunately, in terms of public image, was his involvement in the affair of the 'poisoned chocolates' which were sent to Farnell and his wife in February 1922. The account in *The Times* suggested that these were a direct response to Farnell's earlier and perhaps arbitrary use of power in 'sending down two Communist undergraduates', as well as his banning 'the proposed performance of Grand Guignol plays'. Whatever him sterile must have been growing. The tragedy of their marriage was the lack of children, which might have increased Soddy's bad temper. As mentioned before, he later blamed himself for this, arguing that his work with radium had made him sterile.[30]

Soddy was welcomed to the Alembic Club, where his election to Senior Membership was proposed by Sidgwick.[31] This club was the successor to the Chemical Club of which Soddy had been a founder-member as an undergraduate in 1899. He was also welcomed by the chemistry subfaculty and met them to plan teaching and lectures. In some ways it must have felt like a home-coming. Several of his undergraduate contemporaries were still at the university or had left and already returned. Of the thirteen members of the chemistry subfaculty photographed in the twenties, at least seven had been undergraduates at much the same time. In addition to this, the university's new commitment to science was emphasised by the recent occupancy of one of the other Dr Lee's Chairs, that of Natural Philosophy (or physics), by F. A. Lindemann, later Lord Cherwell.

Within two years of arriving at Oxford, Soddy was awarded the Nobel Prize for Chemistry, the greatest honour for any scientist, and this must have persuaded him that his position was secure, a view supported by the accolades he received from all over the world. Although awarded in 1921, the Prize was not presented until 1922. He had been proposed by Rutherford, who had conveniently been awarded his Prize for chemistry and so was able to recommend a chemist, and seconded by J. J. Thomson.[32] At the presentation ceremony Soddy gave an account of his discovery and paid tribute to all those whose work he felt had contributed to his work, and especially to Alexander Fleck for his tedious but accurate measurements of the beta emitters which made possible the experimental proof of the theoretical proposition. There were well deserved celebrations and congratulations.[33] Many scientists wrote to congratulate him, including Madame Curie, Niels Bohr and Rutherford, and so did H. G. Wells. There was the Nobel Prize banquet with two hundred diners including representatives of the Swedish royal family and the British Ambassador and then a smaller, but possibly just as gratifying, dinner at Merton. This was a more friendly affair. At Soddy's old college various friends appeared, including his friend from his schooldays, Harold Carpenter. Rutherford proposed the toast, and Harold Hartley from Balliol was also present.[34]

There seems to have been only one group missing from the celebrations. Few of his colleagues from Oxford were involved, and this might reflect the extent to which Soddy was already experiencing problems there. These were generally petty, but were all irksome. Because he had not taken his MA, he was not eligible for membership of the board of the

There were somewhat curious circumstances associated with the appointment, which led to problems. Unlike his applications for other posts, successful and unsuccessful, there are no testimonials for this post amongst Soddy's papers, although there are a number there for other posts. Nor is there any other evidence that he applied. A possibly apocryphal explanation for this is that Soddy did not apply but, when the Board of Electors could not agree on the merits of two other candidates, Harold Hartley and Nevil Sidgwick, they agreed to approach a compromise candidate and Soddy was chosen.[25] However, as Davies points out, Soddy had a much better publication and research record than either Hartley or Sidgwick at this point, and whatever the judgement of history on this work, to contemporaries, Soddy must have seemed more eminent.[26] It is also suggested by Dr Collie, Soddy's 'demonstrator for two years, and one of the few Oxford physical chemists who thought highly of him'[27] that the electors 'must have known that he was going to get a Nobel Prize, these things being usually freely discussed a few years before.'[28]

Whatever the truth about his appointment, Soddy's move to Oxford might have benefited from further consideration. In Aberdeen, as a professor at the university, he was a person of considerable importance. He had built up a small but loyal and enthusiastic band of students and had found an audience for his thoughts about the social responsibility of science. The end of the war would have increased his opportunities for developing both these areas. He had a pleasant social life, largely due to the efforts of Winifred, and they were not too far from her parents and the cities of her early life. In addition, the climate, both of weather and of support and opinion, was more attractive in the Scottish setting than in the flat, warm middle of England. All this was changed by the move to a much larger and self-consciously famous university where any alteration to teaching or research structures needed diplomatic dealings with a labyrinthine bureaucratic structure, and where science was still held in particularly low esteem.

On arrival however, the portents seemed auspicious. Soddy and Winifred moved into a charming and spacious house at 131, Banbury Road, one of the areas of North Oxford constructed by speculative building specifically for the university dons and their families. This house gave them space for entertaining, and they set up the weekly social events which had marked their Aberdeen life. Winifred soon found a range of interests in the 'Wives' Guild' and charitable works. She set aside part of the garden to grow plants and flowers, sewed and knitted various objects for Red Cross sales and continued with her painting and sketching.[29] They only needed children to complete their domestic happiness, but, as by this time Winifred was in her mid-thirties and they had been married for twelve years, Soddy's suspicion that radioactive rays had rendered an anaesthetic. The point of Soddy's complaint seemed to be that, during the war, a scheme had been set up through which scientific workers were persuaded to pass discoveries to the Royal Society. These were then passed to manufacturers in exchange for their co-operation during the war and their undertaking 'to prevent these workers from protecting their discoveries from being so stolen'. He was concerned that scientific workers would be deprived of receiving 'due and proper remuneration for the use of their inventions by the government' and that, in turn, he had personally deprived workers in his laboratories in Aberdeen by signing the Royal Society's Declaration on their behalf.[21] Despite Soddy's pleading, the result of the case was 'no award' and must have contributed to his later lack of faith in the Royal Society.

The reasons for his leaving the union were never made clear, but two sets of circumstances provide a possible explanation, both leading him away from scientific affairs altogether. Firstly, he was beginning a lifelong search for an explanation for what he perceived as the ills of his society. To this end he had embarked on a programme of reading in economics which took him away from scientific concerns. Amongst others he read Marx, Ruskin and various theories of credit. The results of this reading are discussed in the next chapter since, although he used his scientific knowledge to construct a 'scientific' monetary theory, based on the laws of thermodynamics, they were not directly relevant to his relationship with Oxford science. Many of his colleagues managed to combine at least two sets of disparate ambitions while holding an Oxford professorship. The case of Lord Cherwell, an exact contemporary of Soddy, which will be discussed below, provides an excellent example of this. For Soddy, however, a second set of circumstances directed his energies away from the union. These were his difficulties associated with his new post.

Soddy's post, the Dr Lee's Professorship, had been created from the Dr Lee's Readership as part of the attempts of the early twentieth century to modernise Oxford science and to provide competition for the Cambridge laboratories and scientists.[22] The post was intended to 'be made thoroughly up-to-date as regards the staff and equipment for teaching and research'.[23] The Professor was to be offered the opportunity to build a radiochemistry research department which would have been comparable with that of Rutherford. If he managed to achieve these aims, Soddy's recognition by the scientific body would be complete.[24] He would also fulfil personal ambitions. Soddy's anxieties throughout his period in Scotland had centred on his isolation from English scientists and on what he believed to be the exclusive attitude of Cambridge scientists. He now had the opportunity of rectifying both of these. These factors provide an explanation, if not a totally satisfactory one, for Soddy's decision to return to Oxford, despite his evident happiness in Scotland.

life is that scientific workers do not exercise in the political and industrial world an influence commensurate with their importance.'[14] While the Cambridge scientists were doubtless important, others also played a part. Soddy, still in Aberdeen, had made his mark on socialist politics through his association with the Union.[15] Although its politics were generally moderate, the Union had a few left wing socialist members who belonged either to the National Guilds League or the Communist Party of Great Britain. Similarly, despite being registered as a trade union, the Union never affiliated to the Trades Union Congress. Initially, the Union was very successful with over a thousand members in the early twenties but the ambiguity expressed in its politics finally destroyed its role as a trade union. In 1927 it became the Association of Scientific Workers, outside the government's definition of a trade union, hoping to attract academic scientists, but never really succeeding.

During the early twenties it fulfilled a function as a body through which science policy could be criticised and evaluated. Its membership included a group of radical scientists including the Professors Soddy and Barstow, Sir William Bragg, the young Hyman Levy and Lancelot Hogben, J. B. S. Haldane and Julian Huxley, as well as H. G. Wells and Sir Richard Gregory, the editor of *Nature*.[16] One aspect of the government's programme for science which they attacked was the role of the DSIR.

Soddy was one of the most active opponents of the role of the DSIR. He believed that 'science—governments science—is being step by step built up; not for humanity but its masters; not for the community, but big business.'[17] At the first meeting of the union in 1920 at Birkbeck College, chaired by H. G. Wells, Soddy presented a more detailed critique:[18]

... the State should have reserved the right to veto researches which were judged to be contrary to the interests of the people. Instead, large sums of money were being handed over for the prosecution of research the results of which were the exclusive property of industry.

In contrast, as *Nature* reported, William Bragg and others were 'more moderate in their views'.[19]

Soddy's close connection with the National Union of Scientific Workers continued through 1920 and into 1921. In November 1920 he was the delegate to the Union from the Oxford and General Branch and was elected President of the Research Council.[20] After this his contacts seem more or less to cease. However, he continued the fight over the ownership of scientific discoveries. In 1923 he was involved in a dispute between the Royal Society and the Commission of Awards to Inventors on one side and King and Mason, scientific workers, on the other. Soddy had become involved as an expert witness but now objected to the way in which the firm had been treated over its wartime production of β-eucaine, into public disagreement with the Government, especially over the vexed question of what was meant by the notion of 'co-operative science'. The hard-learned lessons of the war had led to a belief that government should increase funding for science, but at the same time retain some control over the direction of scientific research, especially promoting research in areas seen as being of importance to industry.[9] This was to be done through the Department of Scientific and Industrial Research (DSIR) and, through this department, Research Associations affiliated to different industries. The Advisory Council of the DSIR was given responsibility for allocating funds to different projects it considered worthwhile. Despite the preponderance of scientific men on the council, problems arose over precisely the relationship between the two branches of science and industry. Industrialists pressed for research into areas of immediate importance to their own concerns while scientists fought to preserve scientific independence and freedom. In addition, there was an important debate over assigning the ownership of scientific inventions. The Government argued that since it paid the salary, laboratory and other expenses of the scientist, it should have rights over both the direction of research and any inventions or discoveries. These questions were important but, as a historian of the DSIR has suggested, 'the tension between freedom and constraint, between planning and the "free market" in research, might not only remain unresolved but be unresolvable.'[10]

Participants on both sides of the debate voiced their concerns. Industrialists worried about both the insistence on sharing research findings and also the lack of freedom in deciding the direction of research. The scientists had other reservations. While the more moderate, such as Sir William Bragg, were grateful for the opportunities which came from this kind of sponsorship, a more radical, and vocal, group voiced their disquiet. Soddy was in the forefront of this group and from 1918 he published his views on the subject. In letters to *Nature* he expressed his anxiety, especially over the question of funding, and argued for this to be put into areas which he regarded as essential to scientific progress, particularly into pure research.[11]

By 1920, his feelings had strengthened. In part, this might have been due to the increasing optimism and resulting moves towards co-operation amongst some scientists which may be perceived during the immediate post-war years. In 1917 the Webbs and the Labour Research Department had suggested that as a part of post-war reconstruction, professional scientific workers would be advantaged by the formation of their own organisations.[12] One result of this was the founding of the short-lived National Union of Scientific Workers.[13] The Union was founded in 1918 by a group of Cambridge scientists whose manifesto argued that 'One of the main reasons why Science does not occupy its proper place in the national other. Professionally he was alone, firstly in Glasgow where he had been appointed to an independent lectureship which was not formally attached to a department and next in Aberdeen because of his war work. In addition, his subject was radio*chemistry* which, for a time, seemed to have no new contribution to make to the science of radioactivity.[3] Perhaps most of all he was set apart by his desire to find a practical application for his discoveries.[4] These various aspects of his life had come together to persuade him to return to England. He had tried unsuccessfully for two earlier posts, the Fellowship at St John's and the Waynflete Professorship. It must also have been galling to hear, early in 1919, that while Rutherford had been appointed to be Head of the Cavendish Laboratories in Cambridge, Soddy's application for a post at Birmingham University had been rejected.[5]

The second consideration affecting his return to Oxford seems to have been his sincere belief that change was possible, and that the spirit of reconstruction felt in almost all areas of society would spread there. He had concluded the remarks quoted above with the question 'is it always to remain a dream, Pygmalion-like, to desire them alive, the brain and heart of the age resident within their walls, and the elements of growth fostered rather than exorcised?' By June 1919, his belief in the possibility of change had strengthened. As he explained to his students at Aberdeen in his 'Farewell Address', this change would come about in two ways.[6] Firstly, and specifically, 'the Universities of Oxford and Cambridge have agreed . . . to co-operate with the Government in setting up Commissions of Enquiry into their affairs'.[7] More generally, he pinned his hopes on the Labour Party which had promised, in its *Report on Reconstruction*, to deal with 'surplus of wealth, which science has created and which is at present absorbed by individual proprietors'. Their aims expressed 'a high ideal totally new in practical politics' and would result in 'the greatly increased public provision that the Labour Party will insist on being made for scientific investigation and original research in every branch of knowledge . . . upon which any real development of civilisation fundamentally depends.'[8] If these programmes were put into operation he could have had high hopes of assisting at the birth of, if not actually founding, a fundamentally new and exciting chemistry department in his old university. He had the opportunity of being near the centre of scientific work with the chance to create his own school and to ensure his reputation at what was generally expected to be the beginning of a new era of hope and prosperity.

However, from the moment of his arrival, Soddy's beliefs in the aims and needs of science were at variance with the Oxford establishment, not only in his allegiance to the Labour Party, but also in his involvement with the National Union of Scientific Workers. These together led him

CHAPTER 5
Oxford and 'big science'

Soddy spent almost the whole of the inter-war period at Oxford. He was appointed the Dr Lee's Professor of Chemistry at Oxford in 1919, a post he held until he retired in 1936. With some reservations, in marked contrast to his years in Scotland, this could be described as a time of almost complete professional unhappiness. Just as his pre-First-World-War years were highly productive, during the inter-war period his scientific work was almost non-existent. However, it was a period in which political and economic work became more and more important to him.[1] For these reasons the inter-war period may be conveniently, if rather artificially, divided into the more or less scientific and the more clearly political areas. Soddy himself would never have recognised this distinction—he always saw himself as a scientist whose science led him to a consideration of matters affecting the functioning of the wider society. However, for clarity, the areas of interest in his life during these decades will be considered in three chapters, the first looking at science, the second at his work on monetary theory in the 1920s and the third at his political involvement in the 1930s.

The first question about Soddy and his time at Oxford must be why he accepted the post. His time as a student had not been especially happy, and his view of Oxford and Cambridge had become more critical throughout the intervening period. In August of 1918 he had drawn up one of his most bitter indictments against them, in which he referred to the way in which 'the old universities remain much as they were, paralysed by the past, and probably even less well disposed to change than they were fifty years ago ... Monuments of bygone days, they remain changeless and resistant as marble, owning no law other than crystallised convention, no logic save that of the stricken blow' and asking 'how it is that the ancient universities can lag so far behind the spirit of the age, and can drag the country with them even to the brink of national extinction.'[2]

Soddy's return to Oxford was based on two separate considerations. The first was his personal ambition. Until this point in his career, Soddy had been something of an outsider. During his time in Scotland he had, geographically, been far from the other centres of research into radioactivity in Manchester and Cambridge, and far from London where meetings of the Royal Society kept scientists in touch with each

53. Varcoe, I. (1970). Scientists, Government and Organised Research in Great Britain 1914–1918: The Early History of the D.S.I.R. *Minerva*, **8**, 200.
54. Quoted in Varcoe, I. (1970). Scientists, Government and Organised Research in Great Britain 1914–1918: The Early History of the D.S.I.R. *Minerva*, **8**, 199.
55. Turner, F. M. (1980). Public Science in Britain, 1880–1919. *Isis*, **71**, 589–608, this quote from p. 602.
56. *Science Progress*, January 1917; Professor Frederick Soddy Papers, Bodleian Library, Oxford, MS Eng Misc b172.
57. Professor Frederick Soddy Papers, Bodleian Library, Oxford, MS Eng Misc b179.
58. Professor Frederick Soddy Papers, Bodleian Library, Oxford, MS Eng Misc b179.
59. Professor Frederick Soddy Papers, Bodleian Library, Oxford, MS Eng Misc b179.
60. Soddy, F. (1920). *Science and Life: Aberdeen Addresses*, Chapters 2 and 4. John Murray.
61. *The Aberdeen Free Press*, 7 April 1915.
62. *Forward*, 2 January 1915.
63. Soddy, F. (1920). *Science and Life: Aberdeen Addresses*, p. 41. John Murray.
64. Soddy, F. (1920). *Science and Life: Aberdeen Addresses*, p. 36. John Murray.
65. Soddy, F. (1920). *Science and Life: Aberdeen Addresses*, pp. 52, 58. John Murray.
66. Soddy, F. (1920). *Science and Life: Aberdeen Addresses*, p. 53. John Murray.
67. Professor Frederick Soddy Papers, Bodleian Library, Oxford, MS Eng Misc b172.
68. Soddy, F. (1918). *Scientific Research*, pp. 7–8. Labour Representative Committee for Scottish Universities, Glasgow.
69. *Daily Herald*, 30 January 1920.
70. *The Lord Rector Aberdeen University Labour Club Magazine*, 19 October 1921, 8–9.
71. *Alma Mater*, the students' magazine of the University of Aberdeen, 9 November 1921, 1.
72. Anderson, R.D. (1988). *The Student Community at Aberdeen 1860–1939*, p. 99. Aberdeen University Press.

31. Edgerton, D. E. H. (1990). Science and War. In *Companion to the History of Modern Science*, (ed. R. C. Colby et al.), p. 940. Routledge.
32. Soddy, F. (1915). The Density of Lead from Ceylon Thorite. *Nature*, **94**, 615.
33. Soddy, F. (1917). The Separation of Isotopes. *Journal of the American Chemical Society*, **39**, 1614.
34. Soddy, F. with Cranston, J. A. and in part with Miss Hitchins (1918). The Parent of Actinium. *Proceedings of the Royal Society*, A **94**, 384.
35. *Pioneer*, p. 203.
36. *Pioneer*, p. 202.
37. See Chapter 2.
38. Marwick, A. (1991). *The Deluge: British Society and the First World War*, pp. 278–9. Macmillan.
39. Varcoe, I. (1970). Scientists, Government and Organised Research in Great Britain 1914–918: The Early History of the D.S.I.R. *Minerva*, **8**, 192–216
40. For the Royal Society see Varcoe, I. (1970). Scientists, Government and Organised Research in Great Britain 1914–1918: The Early History of the D.S.I.R. *Minerva*, **8**, 200. For the politics of these organisations see Werskey, P. (1971). British Scientists and 'Outsider' Politics, 1931–1945. *Science Studies*, **1**, 67–83; Edgerton, D. E. H. (1990). Science and War. In *Companion to the History of Modern Science*, (ed. R. C. Colby et al.), pp. 934–45. Routledge.
41. Quoted in Poole, J. B. and Andrews, K. (ed.) (1972). *The Government of Science in Britain*, pp. 61–3. Weidenfield and Nicolson.
42. Varcoe, I. (1970). Scientists, Government and Organised Research in Great Britain 1914–1918: The Early History of the D.S.I.R. *Minerva*, **8**, 207.
43. Hackmann, W. (1984). *Seek and Strike: Sonar, Anti-submarine Warfare and the Royal Navy 1914–1954*. HMSO. Chapter 2 contains a detailed account of the BIR during this period, relying heavily on Macleod, R. M. and Andrews, E. K. (1971). Scientific Advice in the War at Sea 1915–1917: The Board of Invention and Research. *Journal of Contemporary History*, **6** (2), 3–40. See also Pattison, M. (1983). Scientists, Inventors and the Military in Britain, 1914–1919: The Munitions Inventions Department. *Social Studies of Science*, **13**, 521–68.
44. Quoted in Hackmann, W. (1984). *Seek and Strike: Sonar, Anti-submarine Warfare and the Royal Navy 1914–1954*, p. 17. HMSO.
45. Page, K. (1979). Frederick Soddy: The Aberdeen Interlude. *Aberdeen University Review*, **162**, 133.
46. Snow, C. P. (1961). *Science and Government*, pp. 27–8. Oxford University Press.
47. Page, K. (1979). Frederick Soddy: The Aberdeen Interlude. *Aberdeen University Review*, **162**, 135.
48. *Pioneer*, pp. 204–5.
49. Haber, L. F. (1986). *The Poisonous Cloud: Chemical Warfare in the First World War*. Clarendon Press, Oxford. This book contains the most recent and detailed account.
50. Page, K. (1979). Frederick Soddy: The Aberdeen Interlude. *Aberdeen University Review*, **162**, 141.
51. Page, K. (1979). Frederick Soddy: The Aberdeen Interlude. *Aberdeen University Review*, **162**, 142.
52. Quoted in *Pioneer*, pp. 274–5.

3. Soddy's diary, 1913–14, notes at back. Professor Frederick Soddy Papers, Bodleian Library, Oxford, MS Eng Misc b170, f11.
4. Rutherford Papers, Cambridge University Library, AD 7653, S173.
5. Fussel, P. (1975). *The Great War and Modern Memory*, p. 9. Oxford University Press.
6. Marwick, A. (1991). *The Deluge: British Society and the First World War*, Chapter 3. Macmillan.
7. Marwick, A. (1991). *The Deluge: British Society and the First World War*, Chapter 2, Part 3. Macmillan.
8. Much of the factual information about the personnel and organisation of the Aberdeen Department comes from Page, K. (1979). Frederick Soddy: The Aberdeen Interlude. *Aberdeen University Review*, 162, 127–48. I have not footnoted this material separately here. The interpretation is my own.
9. Professor Frederick Soddy Papers, Bodleian Library, Oxford, MS Eng Misc b178.
10. Page, K. (1979). Frederick Soddy: The Aberdeen Interlude. *Aberdeen University Review*, 162, 131.
11. *The Aberdeen Daily Journal*, Friday 16 October 1914, p. 3.
12. Quoted in Easlea, B. (1983). *Fathering the Unthinkable: Masculinity, Scientists and the Nuclear Arms Race*, p. 52. Pluto Press.
13. Wells, H. G. (1914). *The World Set Free*. Macmillan.
14. Smith, D. C. (1986). *H.G. Wells: Desperately Mortal*, pp. 83–4. Yale University Press, Newhaven.
15. Wells in a letter to Simmons. Quoted by Winifred Simmons in: H.G. Wells as Atomic Seer. *Morning Herald* (Sydney, Australia), 23 February 1946, p. 6.
16. Wells, H. G. (1914). *The World Set Free*, p. 31. Macmillan.
17. Wells, H. G. (1914). *The World Set Free*, p. 31. Macmillan.
18. Wells, H. G. (1914). *The World Set Free*, p. 46. Macmillan.
19. Dickson, L. (1969). *H.G. Wells; His Turbulent Life and Times*, p. 229. Macmillan.
20. Easlea, B. (1983). *Fathering the Unthinkable: Masculinity, Scientists and the Nuclear Arms Race*, p. 55. Pluto Press.
21. Easlea, B. (1983). *Fathering the Unthinkable: Masculinity, Scientists and the Nuclear Arms Race*, pp. 101–2. Pluto Press.
22. Smith, D. C. (1986). *H. G. Wells: Desperately Mortal*, p. 85. Yale University Press, Newhaven.
23. *The Times*, 9 August 1945, p. 5.
24. *Pioneer*, p. 193.
25. Soddy, F. (1903). Recent Advances in Radioactivity. *Contemporary Review*, May 1903, 720.
26. Chapter 3 above; Soddy, F. (1909). *The Interpretation of Radium*. John Murray.
27. Soddy, F. (1912). Transmutation, the Vital Problem of the Future. *Scientia* 11, 182–202.
28. Easlea, B. (1983). *Fathering the Unthinkable: Masculinity, Scientists and the Nuclear Arms Race*, p. 52. Pluto Press.
29. Marwick, A. (1991). *The Deluge: British Society and the First World War*, p. 92. Macmillan.
30. Marwick, A. (1991). *The Deluge: British Society and the First World War*, p. 92. Macmillan.

Either individualism must give way to Socialism and Co-operation or Science must stop . . .

Science is an actual working socialism, communistic in its inheritance and communistic in the spirit of its application. Common ownership of the acquisitions of science is the only path of progress, the only way in which the sum total of human happiness can be augmented.

That is why I, as a scientific man, endorse the aspirations of the Labour movement of today: it alone stands for any ideal above the tawdry and offers an escape from the evil legacy of the old unhappy days.

In his fight for the Rectorship, Soddy was supported by H. G. Wells, George Smillie, the Miners' Leader, Miss A. Maude Royden, Preacher at the City Temple, London, J. C. Squire, the editor of the *London Mercury*, R. H. Tawney and G. D. H. Cole. The latter sent a longish testimonial which suggested that he knew Soddy, saying, 'He has not only academic distinction, but also a clear view of the social problems which most of our Universities are signally failing to deal with. He has realised that professional men need Trade Union organisation fully as much as manual workers, and has taken a prominent part in the organisation of the National Union of Scientific Workers.'[70] Despite this impressive array of support and the hardworking efforts of the University Labour Club, who even succeeded in getting Einstein to comment on Soddy's scientific excellence, the odds against a Labour Candidate were just too great. Despite a 'remarkably plucky show',[71] Frederick Soddy came third, 'behind the Asquithian Liberal Sir Donald Maclean and the victorious Conservative, Sir Robert Horne, a minister in the Coalition Government'.[72]

Soddy's period at Aberdeen had been of fundamental importance to him not so much because of his scientific achievements, which were arguably slight, but because of the transformation in his personal and political values. These, although in part formed by his years in Glasgow, were still unclear in 1914. A combination of the events of the Great War, and especially the use of science during its course, pushed him into a more formulated social critique. Briefly, as it turned out, this took the form of a commitment to socialism. In the longer term it marked the beginning of the change from Soddy as a respectable member of the scientific establishment to Soddy the social critic and 'outsider'.

Notes and references

1. Professor Frederick Soddy Papers, Bodleian Library, Oxford, MS Eng Misc b170, f4.
2. Notes in back of laboratory notebook, dated 29 September 1919. Professor Frederick Soddy Papers, Bodleian Library, Oxford, MS Eng Misc b186.

better advantage than we do.'[65] In view of his feelings about science and war described above, the notion that the enemy might have a scientific advantage was particularly terrifying. However, instead of using this as the basis of anti-German propaganda, he stressed the centrality of science, properly controlled, to peace and reconstruction in all countries. He advocated universal higher education and 'the enrichment of the life of the common people and the elevation of ideals'.[66]

Soddy's identification with the emerging Labour Party became clearer early in 1918, when he wrote an eight page pamphlet, *Scientific Research*, for the Labour Representation Committee for Scottish Universities.[67] In this he expanded at some length on one of his favourite subjects, the advantages which scientific discoveries have brought to society, especially through 'the steam-engine, the energy of coal, fuel and waterfall'. In this, he was saying little more than the Reconstruction Ministry, but he went further: 'If the present achievement is great, the promise of the future is limitless and incalculable.' However, and it is worth quoting his conclusion at length,[68]

All this increment of wealth has not benefited its creators nor the public at large ... Science has come to be for the people little better than a bye-word for commercialism, materialism and greed, for millionaires and slums, polluted rivers, smoky cities and desecrated landscapes, the economic submergence of the unfit, concentration of political power in the hands of the corrupt, an agency for working alternatively on the fears and cupidity of nations, its creative promise prostituted to the work of destruction, and the whole of the wealth accumulated by its exploitation flung like a gambler's stake into the furnace of war ... The hope of the world rests with warm hearts and clear heads, hearts that beat in sympathy with the mass of its workers, and heads that are illumined with the light of science, the science which, pursued solely in the single-eyed service of truth, nevertheless has put within the reach of men the power to re-mould their destinies nearer to the hearts desire.

Soon after this, Soddy moved to a professorship at Oxford, but this did not totally sever his connections with Aberdeen, nor did his politics change direction, in fact rather the reverse. He stood as Labour Rectorial Candidate for Aberdeen University in 1920. In the *Daily Herald* in January of that year he announced his complete conversion to socialism, even if it remained his own particular form, in a front page interview, in which he related his beliefs about science to his commitment to the Labour Party:[69]

The uses already made of science show how necessary it is that a new social order be developed before a million times more awful powers are unleashed by man ...
   The ideals for which the Labour movement stands are the only ones under which the further great gifts of science can be safely entrusted to the world ...

in physical science', and in 1918 on 'Science and human welfare.'[58] He gave another talk in the same year on 'The social abuses of science'.[59]

While this list demonstrates the development of one aspect of Soddy's personal politics, it probably underestimates the extent to which what may be called his non-scientific political and social development was accelerating throughout the period. As described above, during his time at Glasgow from 1904–14, he had become involved through the Beilbys in the suffragette movement and politics of a New Liberal or Fabian kind, although quite how deep his involvement was, especially in the politics, is unclear. As already discussed, his writings suggest echoes of Morrisite utopianism which are likely to have come from socialist contacts in the city. In Aberdeen he seems to have moved more firmly to the left. In 1915 and 1916 he lectured to the Aberdeen branch of the ILP on 'Physical Force' and on 'Science and the State'.[60] It is important to stress that such a public identification could not have been either arbitrary or lightly taken. Although the Aberdeen branch may have been atypical in 1915, the ILP was openly anti-war. As *The Aberdeen Free Press* in the spring of that year put it, the party's politics were 'the one jarring note in an unexampled display of national unity'.[61] When John Bruce Glasier visited Scotland in early 1915 he had no doubt that all Scottish branches were 'sound as a bell and firm as rock . . . [in] their opposition to Militarism in all its forms'.[62] By 1916, when Soddy was still lecturing to the ILP branch, the party was effectively banned in many areas of Scotland and several of its members had been prosecuted for distributing treasonous leaflets.

Soddy's message to the ILP in 1915 was complicated. He did not deliver, as many who spoke at such meetings would have done, simply an anti-war line. He argued that German militarism was, and had been, a threat to all other forms of government. This war was not, he said, 'a war between irreconcilable principles. It is a war between the fundamental principle of all national co-existence and its contemptuous negation.'[63] It was therefore necessary to pursue it as best we could. However, much else in this lecture followed on from his inaugural speech. Science had now made war more terrible than previous generations could have believed, but the future could be worse. Atomic power might be harnessed and used, like nitrogen fixation, for bombs or other military use. He said, 'Imagine, if you can, what the present war would be like if such an explosive had actually been discovered instead of being still in the keeping of the future . . . Surely it will not need this last actual demonstration to convince the world that it is doomed, if it fools with the achievements of science as it has fooled too long in the past.'[64] In his 1916 lecture he returned to his concern about German science and its application to war. He referred to 'the initial supremacy in military science of the enemy' and described the way in which 'Germany uses her people to infinitely role than this and that through its imperial connections and emphasis on 'efficiency' it became 'a conservative, social imperialist pressure group seeking to combine the intellectual prestige of science with the political attraction of efficiency and empire'.[55] However, in its early days the Guild was prepared to support Soddy, as a scientist, against at least a small part of the establishment.

During 1916 Soddy became aware of the terms of the Carnegie Trust for the Universities of Scotland and interpreted his findings to mean that, while the Trust had been set up especially to benefit science, its terms had been interpreted in such a way that a large proportion of the money had been diverted into the humanities. In January 1917 he published 'A criticism of the financial operations of the Carnegie Trust for the Universities of Scotland' in *Science Progress*.[56] This resulted in an investigation by a committee of the British Science Guild. In December 1917 the committee published its findings which largely supported Soddy. However, the trustees disagreed. The result was another debate which lasted into 1918 without any agreement being reached, and contributed to Soddy's general feeling of dissatisfaction at the end of the war.

In general his concerns were more to do with the way in which the authorities were co-ordinating the war effort, science and the future peace than they were with the question of whether science should be used in the war. Just as in *The World Set Free*, a period of world war demonstrated the futility of using science for warfare before the remaining population could put science to peaceful and productive ends, so the lessons of this war should be put to good use. In Wells's post-war fictional world all social problems were solved by the benefits brought about by scientific discoveries and in particular by the energy from harnessing atomic energy and similarly, as his contribution to the debates about the use and responsibility of science show, Soddy felt that it was vitally important that scientific research should continue towards a new and improved post-war world. At the same time, it was important that social organisation should be such that, whatever authorities governed society, they should be aware of potential benefits and so allow free circulation of the results of scientific progress.

Soddy's development of his notions of the social responsibility of science is illustrated by a number of public lectures which he gave to lay audiences and to the students in Aberdeen, some of which he collected and published as *Science and Life* in 1920, others of which remained unpublished. They all shared a common theme, that of the social responsibility of science, or the role of science in society, as their titles make clear. In 1914 he had lectured to science teachers on 'The influences of scientific advances upon life'.[57] He gave two lectures to the Aberdeen WEA, in 1915 on 'The social effects of recent advances to be no instance of scientists arguing unequivocally for the war even when it could be suggested that it contributed to scientific research. The price to be paid in terms of human life and misery was just too high. However, there was a general consensus that the war needed to be fought and that once started, all resources should be used towards a speedy resolution. This was essentially Soddy's position. As his letter at the end of the alkali boiler affair in particular demonstrates, Soddy was not against the war, nor was he against the use of science in that war once it had been begun. What he, like many others, regretted, was that discoveries like the fixation of nitrogen, which in peacetime could contribute so much to the well-being of society, could be put to such 'evil' ends as fighting. However, whenever a choice needed to be made, he chose to fight Germany with whatever powers were available.

It is interesting to note here that Soddy himself interpreted his reactions to the war very differently in his memoirs, written some forty years later. He was quite clear that the war had a fundamental effect on his beliefs, turning him away from science towards some kind of improvement of society, and that this stemmed from one particular event, the death of Moseley. Soddy later remembered that his reaction to the news was extreme:[52]

I think that something snapped in my brain—something which divorced the present from the past. I felt that governments and politicians, or man in general, was not fitted to use science—which obviously he did not comprehend. My mind turned from scientific research to finding out the reason for the failure of science to benefit mankind and bring peace and plenty.

This statement has been taken at face value to explain his subsequent turning away from science towards social and economic matters. However, this is to oversimplify. Moseley was killed in 1915 and Soddy continued his scientific work until at least 1919. His attempts to find the reason for 'the failure of science' had pre-dated the outbreak of war and had continued throughout, as a glance at his publications and lectures throughout this period shows.

This view is supported by a particular episode during the war which, considered alone, appears to be merely an example of petulance. As early as 1905, the British Science Guild had been founded by Sir Norman Lockyer to replace the largely ineffectual British Association for the Advancement of Science of which he had been President.[53] The founding of the Guild had had the powerful support of *Nature*, the most influential British journal for science, and its assistant editor, Richard Gregory, a socialist, helped start the *Journal of the British Science Guild* in 1915. The purpose of the Guild was for 'applying scientific methods to public affairs'.[54] It has since been argued that the Guild had a rather wider communications than the BIR. At first, research into defences against gas had been under the control of a Royal Society Physiology (War) Committee and the Army Hygiene Laboratory. Work on the offensive uses was, from June 1915, under the newly formed Ministry of Munitions, in which Beilby was again on the central panel. This led to the most enormous organisational difficulties; the separation of duties was never formalised; contact between the two bodies was minimal; the initial response of these bodies to any problem was to form a subcommittee. At the same time, the Ministry of Munitions grew in complexity as different responsibilities were added to its remit.

Not surprisingly, research was slow and little was achieved and, in March 1918, there was still a shortage of the chemicals needed to manufacture mustard gas, especially ethylene. Soddy was to investigate its extraction from coal and coke oven gases using charcoal to absorb the ethylene which could then easily be expelled again. The project took up all of Soddy's spare time in the first half of that year. He continued his teaching but everything else was sacrificed to the new project and such was his progress that he applied for a patent in May. The process was never actually used, but led to yet another confrontation with (this time) the Ministry of Munitions. By September 1918, Soddy felt that others in the Ministry were developing his ideas, or ideas which were almost indistinguishable from his. He wrote urgently to General Moulton, the Director General of Explosive Supplies, enquiring about the matter and his status. Eventually he was interviewed by the General who explained that he was opposed to individuals taking out patents for methods developed for the department. Soddy's answer was that he would prefer to pay for his researches than to be regarded as an employee of the Department. He seems to have repeated this view in rather stronger language. Unfortunately there is no existing copy of the letter, but the reply was that 'I am in receipt of your letter but its tone renders it impossible for further notice to be taken of its contents'.[50]

Throughout the whole of the war, Soddy's industry deserves praise, even when the results were less than he would have hoped. However, it has been suggested that there is an ambivalence about his motivation. Kenneth Page, in his account of Soddy at Aberdeen, says, 'Soddy was in a somewhat equivocal position over the misuse of science. Whilst decrying the abuse of science in war, he showed no reluctance in putting his own services at the disposal of the establishment.'[51] He points to the 1918 work on mustard gas in particular to support this interpretation. While Soddy's actions can be read in this way, this seems too simple. Firstly, the ambiguity of his position was not unusual. As suggested above, even Rutherford, the least imaginative of men outside his laboratory, had doubts about the use of scientific discoveries in war and there seems they made 'an effort of indoctrination in the services (and a mutual give and take between serving officers and scientists) . . .' which led to '. . . the smoothness, the lack of friction, and the effortless speed which can only happen in England when the Establishment is behind one'.[46] Unfortunately, in the case of the BIR, the lack of this kind of trust and Admiralty control over the scientists led to a degree of resentment which was to result in the disintegration of the Board and its replacement by a Director of Experiment and Research and a Central Naval Research Establishment.

For the individuals involved, the lack of communication and trust brought about by the tensions between the Board and the Admiralty could result in frustration and annoyance as well as time wasted through duplication of research in just the ways the Board was intended to prevent. Thus although Soddy might have been expected to fare rather better than most since he knew many of those involved with the BIR, he found that he had to repeat some of Carpenter's work on alloys and then from the autumn of 1917 until February 1918 he heard nothing at all and his letters were unanswered.[47] In his letters, he explained that he 'deprecate[d] putting off the real construction till the end of the war. It does not seem to me that the best way to beat a scientific enemy like Germany, simply to produce masses of pre-war weapons . . .' and when this produced no result, wrote that 'it is very regrettable that detailed efforts to answer a problem asked by the Admiralty themselves should be dealt with in this contemptuous and insane manner.'[48] There is no record of any answers to these letters and they had no effect, the work was never finished and Soddy never published his results.

The whole episode reawakened his distrust of authorities, a distrust which had been quiescent since his arrival at Aberdeen, but despite this, in the spring of 1918, Soddy agreed to a request from Lieutenant Colonel Hall of the Explosives Department at the Ministry of Munitions to investigate the extraction of ethylene from coal and coke oven gases. The ethylene was needed for the manufacture of mustard gas, at the time seen as one of the most urgent scientific tasks related to the war. The allies had never really caught up on the German advantages in chemical manufacturing despite mobilisation of most university laboratories and scientists, and one of the areas in which this was particularly felt was in the manufacture of poison gases. Much has now been written about the ineffectiveness of gas as a weapon, but at the time the unreasoning fear engendered by the mere suspicion of gas attacks made it an essential part of the armoury of the forces on both sides.[49] As a result, organisation of research into gas manufacture and anti-gas substances was seen as a priority but, unfortunately, the organisation set up to deal with it was even more cumbersome and prone to disagreements and breakdown in

Each side must be perpetually producing new devices surprising and outwitting its opponents. Since the war began the German methods of fighting have changed time and time again . . . On our side we have not so far produced any novelty at all except in the field of recruiting posters . . . We have produced no counter-stroke at all to the enemy's submarines . . .

The concerns over lack of progress and increasing military difficulties led to the consideration of various government initiatives, and a number of boards were set up to ensure that research was directed towards, firstly, matters of military significance and secondly, the industrial base both in wartime and afterwards. The most influential of these was to be the Department of Scientific and Industrial Research which was formed in 1916 and continued for nearly fifty years. It grew out of a Committee of Council formed in July 1915 which was 'responsible for the expenditure of any new moneys provided by Parliament for scientific and industrial research', and was to operate under the Boards of Education with a smaller Advisory Council formed to advise on scientific and industrial matters.[42]

Of less overall importance, but more relevance to Soddy, two boards were formed to co-ordinate military and civilian research.[43] They were the Munitions Inventions Department, related especially to the Army, and the 'B.I.R.', the Admiralty Board of Invention and Research, later known as the 'Board of Intrigue and Revenge'. The problems it was intended to solve were described by Admiral Lord Fisher, the first chairman of the BIR, in his letter of acceptance of the post:[44]

Man invents: monkeys imitate. The war is going to be won by inventions. Eleven months of war have shown us simply as servile copyists of the Germans. When they have brought explosive shells into damnable prominence, then so have we . . . Noxious gases made us send professors to study German asphyxiation! German mines and submarines have walked ahead of us by leaps and bounds . . .

The Board was deliberately set up to be independent of the Admiralty and a number of civilian scientists were co-opted to serve on its various panels. George, now Sir George, Beilby was a member of the advisory panel, as he had already made the acquaintance of Fisher in 1912 while working on a committee concerned with the use of oil fuel in warships.[45] William Bragg and Rutherford were on a consulting panel, and Harold Carpenter was on the metallurgy panel. This should have guaranteed a level of excellence in the direction of its work, which could be compared with a rather rosy account of a similar organisation, the Tizard Committee, written by the partisan C. P. Snow. This later committee, composed of civilian scientists, but dedicated to assessing the direction of research towards military involvement in a probable imminent war, was successful because with his wife's father, George Beilby.[37] Harold, now Professor, Carpenter, the schoolfriend who had first introduced Soddy to Oxford, and a friend of Beilby, was now at the Royal School of Mines and was to develop alloys which would resist corrosion within the boiler. With such good friends working on the same project this should have been a congenial and successful undertaking. At first it seemed that it would be. Work began on this in January 1916 and continued until February 1918 and at first it went well. Then problems arose co-ordinating the contributions by Soddy and Carpenter, there were difficulties about resources to pay for the more expensive stages of experimentation. Most regrettably, this whole affair demonstrated weaknesses in the way in which research was organised in England and especially, during the war, under the Admiralty Board of Invention and Research.

Although given extra urgency by the whole question of the war, the problem of the organisation of science was one which had been exercising various groups throughout the early twentieth century. One of the problems was a conflict of ideologies, between the ideals of teamwork and organisations promulgated by the government and the particular scientists. It has been suggested that 'the advancement of science contributed to the collectivist idea' and there is certainly some evidence to support this view in the formation of the National Union of Scientific Workers in 1917 and in policies for reconstruction, especially those of the Labour Party, in which the emphasis was on the importance of a scientific society, in approach as well as research.[38] However, the scientists were individualistic, accustomed to working in a highly hierarchical, competitive system within laboratories and institutions and, once reaching the top of their ladder, tending to conservatism, wanting to perpetuate a system through which they had struggled. The latter tendencies were most marked in wartime government research policies.[39] It seems clear that the boards set up to implement the wartime policies not only recruited from a very limited pool, so that a number of individuals sat on more than one board, but they also shared at best a conservative, reformist attitude to the whole question of the reform of scientific research. Their views were largely shared by the Royal Society although there were differences of opinion about which organisation should have control of any funds and therefore power over how public science and research could be established. The Royal Society as a body was of the opinion that it should control all science and committees since it consisted, by definition, of most of the senior research scientists.[40]

In view of these differences, it is perhaps unsurprising that the establishment of any policies during the early days of the war appears to have had little effect on the progress of organising scientific research. As H. G. Wells put it in a letter to *The Times* in June 1915,[41] were appointed as assistants, making history within the university in the process, but again reflecting the national experience.

The results of these changes were that undergraduate teaching was just about covered but the other side of the department's work, research, was difficult. Until the departure of Miss Hitchins and Cranston, one to undertake war work of an unspecified kind and the other to enlist, Soddy worked in 1914 and 1915 on the parent nuclide of actinium, on the half-life of ionium and the properties of thorite lead with them, both of whom had come from Glasgow with him. Apart from a short period in 1917, this was the only 'pure research' undertaken during the war and resulted in the publication of only two scientific papers. Soddy's publication of scientific works during his time at Aberdeen demonstrates the obstacles with which he was working. The publications consisted of a series of Royal Institution lectures on radioactivity which he gave in 1915, 'The density of lead from Ceylon thorite', which was published in *Nature* in the same year and refers to some of his work on an isotope of lead found in Ceylon thorite.[32] This was a part of his long term concern to find an alternative source of radioactivity to radium. In 1917 he published 'The separation of isotopes' in America[33] and lectured at the Royal Institution on 'The Complexity of the Chemical Elements'. In 1918 there was, with Cranston and Miss Hitchins, 'The parent of actinium'.[34] Although these would seem to represent a respectable output, by Soddy's own standards there was actually little original work involved and in fact these were to be the last research papers which he published. At the time it seemed too that there was just a temporary pause in reaction to the exigencies of war since the amount of work for the military authorities was all but overwhelming.

The laboratories were involved in war work of various kinds for the duration of the hostilities. Soddy's organisational ability was exercised to the full when the Sectional Committee of Chemistry of the Royal Society requested that the chemistry department should undertake the synthesis of the chemicals required for the manufacture of Novocain and other local anaesthetics which had also been manufactured in Germany. Under Soddy's direction the laboratories were turned into a minor factory and, despite various difficulties, the chemicals were produced by day and night shifts. In addition to this a number of enquiries from the Royal Aircraft Factory were answered; the Optical Department of the Ministry of Munitions and the Compass Department of the Royal Navy consulted him about the possible uses of radium and 'A proposal to employ a certain substance in gas warfare was reported on'.[35]

The major project of the war for Soddy was research into an alkali boiler to be used in submarines 'as an alternative for electric propulsion under water'.[36] This project can be seen as an almost direct continuation of the work Soddy had done on batteries, fuels and similar projects in Glasgow industries. Attempts were made to reverse this during 1915, but for some, the pull of the recruiting campaign had been too strong to resist. One of these was Rutherford's most brilliant research student, H. G. J. Moseley, who enlisted in the army as a signals officer. His experience has been used to show how talent could be wasted in the armed forces where there was no clearly defined route for a scientist. At the time, his enlistment and that of others like Haldane caused a section of the established scientific world, led by Rutherford, to voice their outrage at this waste, and to demand that such men should be found posts which would utilise their skills. This outrage was an emotional rather than a considered reaction since it later emerged that Moseley had been offered scientific work which he refused,[30] possibly seeking 'like many other middle-class Europeans . . . to escape the humdrum materialism of everyday life'.[31] Unfortunately, Moseley was sent to the Dardenelles where he was killed in April 1915.

Others felt the same as Moseley. The early effects of the war had been minimal. Over the university as a whole, there had been a reduction in student numbers of about a quarter, which represented a substantial number of male students being mobilised on the outbreak of war. Two of the staff of the chemistry department left. Soddy himself joined the First Battalion of the City of Aberdeen Volunteer Reserve as a private. None of this had affected the day-to-day life of the department to any great extent. However, a map of 'The Great Battle in Poland', publicity for 'Motor Ambulances for the Front', the 'Belgian Relief Fund', leading inexorably to 'Recruiting at Aberdeen' where 'there is no falling-off in the numbers coming forward for the military and naval forces at Aberdeen' were all to be found on the page of the local paper which reported Soddy's inaugural lecture. Despite attempts to ensure the continuation of staffing levels in certain vital occupations, which included research in laboratories like Soddy's which were soon asked to turn their attention to military concerns, the various pressures on young men (and some young women) brought about the departure of all the research students from Soddy's department for various kinds of war work by the autumn of 1916. By this point, in addition, undergraduate student numbers continued to fall and the proportion of women students to rise. Miss Hitchins went against the national trend when she left the laboratory for different kinds of work; more generally, women replaced men in this, as in so many other areas. However, the replacement could not be complete and by 1917, student numbers in the chemistry class had dropped from about forty to only twelve, of whom eight were female. The war also exacerbated the difficulties of staffing the department. Some of the junior men enlisted, others retired and one of the lecturers was seriously ill. To overcome this, a lecturer from the Technical College was brought in and three women wide public which rapaciously consumed much of Wells's works, Soddy's lecture was a success. His student, Cranston, who had followed him from Glasgow to Aberdeen, described it as 'the great event that stands out in my memory . . . I was present and deeply moved by it'.[24] It also marked a turning point in Soddy's career. Until this point in his life, although he had been involved in controversies of various kinds, he was still well within the mainstream of scientific research and theories of its applications. If some of his lectures were perhaps seen as far-fetched, they were still acceptable to a wide audience—whether the public in Glasgow or the professionals of the Royal Society or the British Association. Before the outbreak of war, as we have seen above, he had developed vague doubts about the power of science over the previous decade. As early as 1903, he had referred to the dangers inherent in the 'gigantic forces' of energy within the universe,[25] and in 1908 he had voiced his increasing fears about the uses of scientific discoveries.[26] Even in his more detailed exposition of 1912, his fears had centred on the depletion of natural resources,[27] but now, in his inaugural lecture, he gave the first public intimation that the fears were more than a general unease about an unspecified future. By referring his audience so explicitly to *The World Set Free* and by endorsing it so fully, Soddy was giving a graphic example of both the potential uses of science to the community and the possible abuses of that science in warfare. His experiences during the war years were to strengthen these anxieties until his views and attitude to science were fundamentally changed. During the same period, Rutherford, despite his sentiments in 1916 quoted above, continued his researches into atomic energy throughout the war and, in 1919, announced that he had achieved the first laboratory transmutation of an element.[28]

This is to anticipate the events of the next four years. By early 1915 the war was becoming more real. 'Business as Usual' had been abandoned 'amid the bangs of the Battle of Neuve Chapelle . . . which jerked society to an awareness of what was involved in this most beastly of wars.'[29] The impact of the war was felt far more strongly throughout mainland Britain from this point onwards and, like the earlier period, this was palely reflected within the chemistry department of the University of Aberdeen. Firstly, there were the problems caused by the naval blockade imposed by Germany from February 1915. The most obvious effect of this was a shortage of foodstuffs, but it had less immediate effects as well. Scientific supplies had previously been obtained from Germany and these were now of course unobtainable. Glass equipment, chemicals and especially radioactive samples were all in short supply or non-existent. Secondly, there was an increasing problem with the supply of labour. Indiscriminate recruiting during the early days of the war had resulted in shortages in essential industries, especially chemical and explosive given them a practical application in the mid-twentieth century. '. . . the problem of inducing radio-activity in the heavier elements and so tapping the internal energy of atoms, was solved by a wonderful combination of induction, intuition and luck by Holsten [the hero of the book] so soon as 1933.'[16] Holsten managed to transmute bismuth into gold—the alchemists dream—but, according to Soddy's prophesy, the importance of the discovery was that it 'opened the way for mankind . . . to worlds of limitless power'.[17] Wells predicted that by the 1950s this power would have replaced all other forms of energy but, 'because the systems of government were so ill-prepared for this, that in the full tide of an incalculable abundance, when everything necessary to satisfy human needs and everything necessary to realise such will and purpose as existed then in human hearts was already at hand, one has still to tell of hardship, famine, anger, confusion, conflict and incoherent suffering.'[18] However, there was worse to come, in the form of 'The Last War'. The difference between this war and previous ones was the kind of bombs, and it is worth quoting Wells's description at some length because it is said to be the first account of a chain reaction.[19] It has been suggested that this account originally inspired Leo Szilard to turn to nuclear physics and subsequently to the discovery of the nuclear chain reaction.[20]

Always before in the development of warfare the shells and rockets fired had been but momentarily explosive, they had gone off in an instant once and for all . . . but Carolinum [Wells's new radioactive element] . . . once its degenerative process had been induced, continued a furious radiation of energy and nothing could arrest it . . . once launched the bomb was absolutely unapproachable and uncontrollable until its forces were nearly exhausted, and from the crater that burst open above it, puffs of heavy incandescent vapour and fragments of viciously punitive rock and mud, saturated with Carolinum, and each a centre of scorching and blistering energy, were flung high and far.[21]

Although the conclusion of the novel was that after the war the few survivors settled to create a technological Utopia where gold was valueless, cities were planned, food produced in laboratories and factories, and the land used for leisure, in typical Wellsian form, there is an underlying ambivalence about this created Eden. He suggests that the science which created it could still have a darker side, especially when set against the grandeur of mountains and nature. The appeal of this to a popular readership in 1914, just as war was about to be declared, was limited, although it reached a small audience throughout the inter-war period. There was a brief revival of interest when it was reissued in 1945 but, not surprisingly, its message still proved too stark for a wide audience.[22] As expressed in a letter to *The Times*, Wells had 'correctly' shown 'the effect and consequences of the use of atomic bombs'.[23]

While reaction to the novel was ambiguous and it never reached the

However, in the conclusion to his lecture, Soddy voiced the darker side of these possibilities. Such 'increase [in] the power of mankind to do work . . . could be used for evil as well as good. An instance of this was to be found in the present war in the use of scientific weapons of destruction.' Although he does not specify which 'weapons of destruction he had in mind' it is likely that he was referring to Fritz Haber's discovery that nitrogen fixation, for which the chemists had recently found a method of synthesis in the laboratory or factory, was useful not only for fertilisers but also for explosives. This was one of the first new scientific discoveries to be used in the war and provides a graphic illustration of Soddy's dilemma. If such an apparently peaceful and productive process as the production of fertilisers for a hungry world could be subverted in such a way, then the problems for atomic power were enormous. Rutherford was to voice similar concerns in 1916 when he declared it fortunate that no method of controlling the release of atomic energy had been found and hoped that 'we should not discover it until Man was living at peace with his neighbours.'[12] However, Rutherford was never to become involved in questions of the politics of the social applications of his discoveries, and as a result was free to continue his researches unhampered by such self-questioning as Soddy was frequently to find himself absorbed in. At this moment though, at the end of his lecture Soddy referred his audience to H. G. Wells's latest work for a demonstration of the issues. He talked of the 'fanciful but perhaps after all not so fanciful' picture of future developments drawn by 'Mr. Wells in his greatest novel, *The World Set Free*'.

It is worth spending a little time looking at H. G. Wells's novel. Since it was based on Soddy's *The Interpretation of Radium*, it was dedicated to Soddy and it reflects so much of Soddy's thinking that it appears as a kind of cipher for his own thinking at that time. *The World Set Free* was a scientific novel which Wells published early in 1914.[13] In 1913 he had, he wrote, 'suddenly broken out into one of the good old scientific romances again. And I suddenly need to know quite the latest about the atomic theory and sources of energy'[14] and so had 'read and mastered Soddy's very good little book and I want more. My idea is taken from Soddy . . .'[15] The result was the dedication of *The World Set Free*, which was

To Frederick Soddy's "Interpretation of Radium":
This story, which owes long passages to the eleventh chapter of that book, acknowledges and inscribes itself.

Wells's discovery of Soddy's book marked the beginning of an acquaintance which was to continue throughout the next two decades and to lead Soddy's thoughts in directions which we will consider below. For various reasons this book was never a success. Wells had taken Soddy's ideas and new course on radioactivity which he taught in the spring of 1915. He instituted a series of lectures on radioactivity which were greeted with the kind of enthusiasm he seems to have been able to instil in any audience whenever he himself was committed to the subject. He had demonstrated the difference between his duties and his personal enthusiasm over and over again, regardless of the constitution of the audience. In Australia he had persuaded miners of the excitement of his subject; in Canada he had almost gone too far—what was near to fanaticism at McGill had inspired his undergraduate audience to the point which had antagonised Rutherford, fortunately only temporarily; in Glasgow he had animated his research students. The difference between his general lectures and those on his speciality was commented on by one of the Aberdeen students. 'They were . . . in quite a different category . . . once he was on his own subject his unique personality revealed itself, and one felt that one was in direct touch with a real discoverer.'[10]

Although he was able to effect little in the way of change in the chemistry department during the first months after his appointment, other less obvious changes were manifesting themselves, especially in his personal philosophy. The direction which these were taking is indicated by some of his points made during his inaugural lecture as Professor of Chemistry at Aberdeen which was given to a 'crowded classroom' on 15 October 1914. It was reported in full in the Aberdeen papers the next day.[11] In his lecture, after the customary acknowledgement of his gratitude at his appointment, Soddy gave an impressive account of how he saw his science and its importance. He brought together in a highly original way what he saw as the important strands in the advancement of chemistry. He described the history of the discipline from the seventeenth century, but, perhaps out of respect to his audience, did not repeat the ideas about the connections with alchemy which he had related to his McGill students. He went on to discuss the connections between the material and non-material world which needed to be considered in any chemical investigation. By 'invisible' in this context he meant, on the one hand, literally those which cannot be seen like the workings of the body and, on the other, the forces of energy. From energy he was able to lead conveniently on to his other great interest, the potential uses of radioactivity. He pointed out that the potential energy contained in radium, which was a million times greater than that in the same weight of coal, would, if only it could be harnessed, provide a cheap and clean source of energy which could be turned into work. This would have an additional advantage of not only being easily available, but also of being a clean fuel. Using this form of energy would mean that the polluting effects of burning coal for energy, seen in the 'ham and eggs cloud' which arose over Glasgow every morning, would disappear.

the war to five feet three.[5] Despite this, shops like Harrods and W. H. Smith used the war as an advertising slogan with their 'Business As Usual' banners, while Parliament attempted to carry on as before even when, in February 1915, the Prime Minister admitted that the price of flour had risen by 75 per cent, imported meat by 12 per cent and coal by 15 per cent. He justified government inaction on the grounds that the rises were no worse than those after the Franco-Prussian war of 1870.[6] In Scotland, especially on Clydeside, the early days of the war were marked by labour unrest, not military fervour.[7]

Soddy's early days at Aberdeen are a pale reflection of these national trends. Just as 'Business As Usual' superficially disguised the initial effects of the war and blanketed tensions associated with the fighting in France and Belgium, nothing changed in Soddy's life or the department he took over. None the less, underneath a tranquil surface pressures were building which were to result in a transformation of both. He had gone to Aberdeen with the intention of setting up a department with a new emphasis on radioactivity, which was after all the whole point of his appointment. However, before any innovation could be attempted, the problems inherent in attempting to take over an established department had to be faced.[8] The most important of these was the staff. The two lecturers, F. W. Gray and J. Knox, and the two assistants, W. H. T. Williamson and J. D. Pratt, had all been students of Soddy's predecessor, Professor Japp, before their appointment. The senior laboratory attendant and his assistant had worked with Japp and there were also five research students already in the department. Against this formidable array of tradition, Soddy had enthusiasm and the arrival of two of his own students, J. A. Cranston and Miss A. F. R. Hitchins, who followed him from Glasgow. Instituting change in the face of this staff would have been problematic, but there was a further legacy from Japp which proved more intractable. The degree structure and the teaching had been put in place during Japp's time and included the requirement that the professor should give one hundred lectures on the principles of organic and inorganic chemistry to undergraduates studying for the First Science Examination, all illustrated with lecture-table demonstrations. While this basic course was no longer at the centre of his interests, Soddy approached the task with his customary carefulness. The surviving series of notebooks show how he wrote out each of his lectures, complete with diagrams for the demonstrations and questions which he should set for the students.[9] He also inherited other tasks. There were other, smaller classes and laboratory sessions to be taken and the postgraduate students to supervise. As a result he was too busy to make any major alterations.

However, he did make some changes: he made the laboratory course compulsory for medical students and, more significantly, introduced a

CHAPTER 4

Aberdeen and the First World War

Winifred and Frederick Soddy arrived in Aberdeen in the summer of 1914 full of hopes for the future. It seemed that Soddy had finally, at the age of thirty-eight, achieved professional recognition. He was professor of a department in a Scottish university and, at least before the First World War, this gave him considerable powers within the institution and respect outside. In addition, it looked as if Aberdeen would suit them very well domestically. For the Soddys, with their love of walking and climbing, its position was very much in its favour since the hills and mountains were easily accessible and there they could escape the city.[1] They arrived prepared to maximise these opportunities. In addition to climbing boots, they took from Glasgow bicycles, golf clubs and a drawing board in anticipation of the enjoyment of their new surroundings.[2] The climate was also an improvement on Glasgow. Soddy had disliked the warm dampness typical of the west of Scotland, although he had suffered less there than he had in London smogs. He much preferred the dryer, colder weather of the eastern and more northern part of the country which was, he found, far more conducive to work.

However, before work or expeditions could be undertaken, the immediate task was to furnish their house at 14, Albyn Terrace. This must have been larger than their Glasgow tenement apartment since Soddy's diary for this year suggests that he spent a good deal of time after their arrival in furniture stores and sales. He made a number of lists of furniture and the appropriate prices: a settee cost £15, a bedroom suite in fumed oak ten guineas and a hall cupboard £7.[3] By the beginning of 1915 they had moved everything safely and furnished their home but were left with one fear, at that point something of a joke, 'Let's hope the German shells will not undo the work'.[4]

In Aberdeen, there was little sign of any effects of the First World War during 1914 or the beginning of 1915. Although war had been declared on 4 August 1914, the policy of 'Business as Usual' on the home front, which had been instituted by a powerful combination of large shopkeepers and the government, was in force. However, the scale of casualties in the first months of fighting was such that, by November, the 'Retreat from Mons', the 'First Battle of Ypres' and the 'Race to the Sea' had decimated the British Army to the point where the standard height for volunteers was lowered from the five feet eight inches it had been at the beginning of

56. Todd, M. (1918). *The Life of Sophia Jex-Blake*, p. 525. Macmillan.
57. Letterbooks of the Glasgow Society for Women's Suffrage, Mitchell Library, Glasgow, p. 702.
58. Todd, M. (1918). *The Life of Sophia Jex-Blake*, p. 525. Macmillan.
59. Letter from Hilda Beilby to Muriel Howorth. Professor Frederick Soddy Papers, Bodleian Library, Oxford, MS Eng Misc b170.
60. Letterbooks of the Glasgow Society for Women's Suffrage, Mitchell Library, Glasgow, p. 703. Information from Mitchell Library librarian, Hazel Wright.
61. Frederick Soddy, letter to *Glasgow Herald*, 30 July 1913, copy in Arncliffe–Sennett Papers, Vol. 25.
62. Professor Frederick Soddy Papers, Bodleian Library, Oxford, MS Eng Misc b179 f170, my emphasis.
63. This has been enough to gain him notice in Raynor-Canham, M. F. and Raynor-Canham, G. W. (1990). Pioneer women in nuclear science. *American Journal of Physics*, 58, 1038. Unfortunately they have missed Miss Hitchins, who was Soddy's assistant for some years, in their research.
64. Arncliffe–Sennett Papers, Vol. 25.
65. Holton, S. S. (1986). *Feminism and Democracy: Women's Suffrage and Reform: Politics in Britain 1900–1918*, pp. 100–70 and *passim*. Cambridge University Press.
66. Rutherford Papers, Cambridge University Library, AD 7653, S131, S136.
67. Soddy/William Bragg correspondence in 6A 38, The William Henry Bragg Collection, Royal Institution.
68. Professor Frederick Soddy Papers, Bodleian Library, Oxford, MS Eng Misc b170, f407.
69. Personal communication from Margaret Shillan.
70. *Pioneer*, p. 99.
71. *Memoirs*, pp. 79–81; *Pioneer*, p. 99; Easlea, B. (1983). *Fathering the Unthinkable: Masculinity, Scientists and the Nuclear Arms Race*, pp. 45–8 and *passim*. Pluto Press.
72. Rutherford Papers, Cambridge University Library, AD 7653, S161–3.
73. Rutherford Papers, Cambridge University Library, AD 7653, S172.

27. Davies, M. (1992). Frederick Soddy: The Scientist as Prophet. *Annals of Science*, **49**, 354.
28. See the article by Krivomazov in Kauffman, G. B. (ed.) (1986). *Frederick Soddy (1877–1956): Early Pioneer in Radiochemistry*, pp. 122–6. Reidel, Dordrecht.
29. Rutherford Papers, Cambridge University Library, AD 7653, S157.
30. For example, see the Soddy/William Bragg correspondence in 6A 24, The William Henry Bragg Collection, Royal Institution.
31. Freedman, M. I. (1986). Frederick Soddy and the Practical Significance of Radioactive Matter. In *Frederick Soddy (1877–1956): Early Pioneer in Radiochemistry*, (ed. G. B. Kauffman), pp. 171–6. Reidel, Dordrecht.
32. *Pioneer*, p. 146.
33. See below.
34. Burchfield, J. D. (1980). Kelvin and the Age of the Earth. In *Darwin to Einstein: Historical Studies of Science and Belief*, (ed. C. Chant and J. Fanvel), pp. 188–90. Longman, Harlow.
35. Fleck, A. (1957). Frederick Soddy. *Biographical Memoirs of the Royal Society*, **3**, 205.
36. Fleck, A. (1957). Frederick Soddy. *Biographical Memoirs of the Royal Society*, **3**, 205.
37. Stothers Glasgow, Lanarkshire and Renfrewshire Annual, **1912**, p. 27.
38. Crosbie-Smith, D. (1989). *Energy and Empire: A Biographical Study of Lord Kelvin, passim*. Cambridge University Press.
39. Smouth, T. C. (1986). *A Century of the Scottish People 1830–1950*, p. 262. Collins.
40. Muir, E. (1940). *The Story and the Fable: An Autobiography*, pp. 117–20. George Harrap and Co. Ltd.
41. Middlemas, R. K. (1965). *The Clydesiders*, p. 31. Hutchinson.
42. Smout, T. C. (1986). *A Century of the Scottish People 1830–1950*, p. 262. Collins.
43. Freedman, M. I. (1986). Frederick Soddy and The Practical Significance of Radioactive Matter. In *Frederick Soddy (1877–1956): Early Pioneer in Radiochemistry*, (ed. G. B. Kauffman), pp. 258–9. Reidel, Dordrecht.
44. *The Compact Edition of the Dictionary of National Biography: Twentieth Century DNB, 1901–1960* (1975), p. 2896. Oxford University Press, Oxford.
45. Professor Frederick Soddy Papers, Bodleian Library, Oxford, MS Eng Misc b179, f103.
46. Burgess, K. (1980). *The Challenge of Labour: Shaping British Society 1850–1930*, pp. 135–6. Croom Helm.
47. Soddy, F. (1912). *Matter and Energy*, p. 35. Williams and Norgate.
48. Soddy, F. (1912). *Matter and Energy*, pp. 247–9. Williams and Norgate.
49. Soddy, F. (1912). *Matter and Energy*, pp. 243–4. Williams and Norgate.
50. Morris, W. (1977). *News From Nowhere*, pp. 6 and 140. Routledge and Kegan Paul.
51. Soddy, F. (1912). *Matter and Energy*, p. 243. Williams and Norgate.
52. Soddy, F. (1912). *Matter and Energy*, p. 244. Williams and Norgate.
53. See the next chapter for a discussion of this book in relation to Soddy and his work.
54. Soddy, F. (1912). *Matter and Energy*, p. 34. Williams and Norgate.
55. See the article by Krivomazov in Kauffman, G. B. (ed.) (1986). *Frederick Soddy (1877–1956): Early Pioneer in Radiochemistry*, p. 121. Reidel, Dordrecht.

Notes and references

1. *Pioneer*, p. 142.
2. This interpretation of Soddy's appointment is fundamentally different from that offered by Muriel Howorth in *Pioneer*. However, as explained in the introduction, I feel that her account is flawed by a heavy reliance on Soddy's own version of events when remembered fifty years after their occurrence and at a time when he was particulary embittered.
3. Glasgow University Court Books, 50551, No. 1.
4. The Glasgow and West of Scotland Technical College. *Stothers Glasgow, Lanarkshire and Renfrewshire Annual*, 1911; Glasgow University Court Books, 50552, No. 2, June 1905 and 50551, No. 1, June 1903.
5. *The Compact Edition of the Dictionary of National Biography: Twentieth Century DNB, 1901–1960* (1975), p. 2509. Oxford University Press, Oxford.
6. Professor Frederick Soddy Papers, Bodleian Library, Oxford, MS Eng Misc b180, f. 66.
7. Glasgow University Court Books, 50551, No. 1, 10 March 1904.
8. Medley's reminiscences [of Glasgow University]. Unpublished manuscript, Glasgow University Archive, No. 827.
9. Russell, A. S. (1956). F. Soddy, Interpreter of Atomic Structure. *Science*, **124**, 1069.
10. Soddy, F. (1909). *The Interpretation of Radium*. John Murray.
11. Fleck, A. (1957). Frederick Soddy. *Biographical Memoirs of the Royal Society*, **3**, 206–7.
12. Rutherford Papers, Cambridge University Library, AD 7653, S130.
13. Rutherford Papers, Cambridge University Library, AD 7653, S131.
14. Rutherford Papers, Cambridge University Library, AD 7653, S133 and S134.
15. Paneth, F. A. (1957). A Tribute to Frederick Soddy. *Nature*, **180**, 1085.
16. For a discussion of the system of recognition and rewards and how it functions within the scientific community, see Barnes, B. (1985). *About Science*, especially pp. 45–9. Basil Blackwell.
17. These remarks are taken from Soddy's letters to Rutherford during the period.
18. See below.
19. Rutherford Papers, Cambridge University Library, AD 7653, S133.
20. Rutherford Papers, Cambridge University Library, AD 7653, S137.
21. Rutherford Papers, Cambridge University Library, AD 7653, S137.
22. Trenn, T. J. (1986). Frederick Soddy. In *Frederick Soddy (1877–1956): Early Pioneer in Radiochemistry*, (ed. G. B. Kauffman), p. xxi. Reidel, Dordrecht.
23. See the article by Krivomazov in Kauffman, G. B. (ed.) (1986). *Frederick Soddy (1877–1956): Early Pioneer in Radiochemistry*, p. 120. Reidel, Dordrecht.
24. This technical account of the discovery of isotopes is essentially that given by Badash, L. (1986). The Suicidal Success of Radiochemistry. In *Frederick Soddy (1877–1956): Early Pioneer in Radiochemistry*, (ed. G. B. Kauffman), pp. 27–41. Reidel, Dordrecht.
25. Badash, L. (ed.) (1969). *Rutherford and Boltwood: Letters on Radioactivity*. Yale University Press, New Haven.
26. Badash, L. (ed.) (1969). *Rutherford and Boltwood: Letters on Radioactivity* p. 32. Yale University Press, New Haven.

an unfailing support to him.[68] The tragedy of their marriage was the lack of children for which Soddy later blamed himself, arguing that his work with radium had made him sterile.[69] It might well have done so. The dangers of radiation were not at all recognised at this stage and so no precautions were taken. By all accounts the Curies' laboratory glowed in the dark and when Soddy bought his radium samples from Isenthal's he carried them back to the university in a glass vial and proudly showed them to Ramsay who took out a sample on a platinum rod, contaminating the whole laboratory in the process.[70] There are accounts of radioactive specimens being sent through the post and Soddy customarily had his experimental samples for lectures in glass tubes on the bench in front of him. It is possibly more surprising that this generation of scientists produced any children and more of them did not succumb to radiation poisoning.[71]

After their honeymoon, Winifred and Frederick settled down to married life in one of the middle-class tenement buildings near to the university and to her parents. They continued to spend holidays with the Beilbys, and often with Harold Carpenter and his wife. They managed to meet the Rutherfords at a couple of conferences and this friendship seems to have strengthened. There were clubs and associations which reflected most of their interests so that Soddy at least attended the Royal Society Dinner, the Ski Club Dinner, meetings of the 'Suffrage Society' and the 'Alchemists' as well as a golf tournament—he had recently learnt to play—during the winter and spring of 1913–14. All this, in addition to his very successful career, should have meant that Soddy was contented, at least for a few years. By this time, as well as his routine work, social and domestic life, and work with Beilby, he had been made a Fellow of the Royal Society, had been awarded the prestigious Cannizzaro Prize and had made his major individual discovery, that of isotopes. However, he still wanted a professorship, and ideally an English one. He had testimonials and a letter of application printed for the post of Waynflete Professor of Chemistry at Oxford in 1912, but when this was unsuccessful he could find nothing else that was suitable.[72]

By the summer of 1914 he needed to decide about his future. His present post was renewable every five years but he was 'not looking forward to another five years'. Ferguson, the senior professor in the chemistry department, was about to complete forty years in his Chair with no sign of retirement. Soddy had to teach increasing numbers of students; his work was increasingly complex and the 'affiliation of the University and Technical College makes the Department more difficult each year.' As a result of all this he had decided to apply for the Professorship of Chemistry at Aberdeen University. He was appointed and by January 1915, he and Winifred were installed in 14, Albyn Terrace, Aberdeen.[73]

I am in general sympathy with the objects of this Federation but I am wondering what is being done in this direction by the existing Suffrage Societies and should have preferred a more united and combined effort on the part of all the societies this autumn which already have large numbers of men on their membership roll.

This is a good reason for doing nothing, and as Mrs Arncliffe-Sennett recognised in a note on the letter 'Might do so much—but won't do anything.'[64]

A contributing factor to this inactivity might have been politics. The Glasgow suffrage societies had been dominated by Liberals since the founding in 1902 of the Glasgow and West of Scotland Women's Franchise Association and this had been a constant cause of friction in a city where politics was increasingly influenced by socialism. This made participation in both party and suffrage politics difficult, and from 1906 there were numbers of resignations from the society through disillusionment with Liberal policies.[65] This would have put Soddy, who seems at this time to have inclined towards the ILP, in a difficult position which was best managed by joining nothing.

It was becoming increasingly important to him that he should control these diverging tendencies because of his relationship with the Beilbys, especially with Winifred Moller Beilby, daughter of Emma and George. Soddy and the Beilbys had more than science and politics in common. Soddy's love of climbing and mountains has already been described and he now found that the Beilbys spent at least one holiday in the mountains each year, sometimes on their way to a conference or meeting, as in the spring of 1906 when they visited Switzerland on their way to the International Congress of Applied Chemistry in Rome; on other occasions just as a holiday, as they did when they returned to Switzerland for Christmas of that year.[66] From 1905, Soddy joined them on these trips. They visited the mountains, walking, climbing, tobogganing and trying the new sport of ski-running. Beilby and Soddy took photographs, and what had started as a largely professional friendship with George Beilby soon became something more personal. Winifred Beilby and Frederick Soddy announced their engagement in September 1906. As he explained to William Bragg, 'she has just come of age and everything has been arranged swimmingly.'[67] They decided not to marry immediately, but waited until 27 March 1908, when they chose to have a very quiet ceremony, early in the morning, before departing on a walking honeymoon. This was the beginning of a relationship which was entirely happy and contented for the next twenty or more years, until Winifred's premature death. All accounts are agreed in praising Winifred. She is noted for her charm as a hostess, she was able to deal with Soddy's increasingly bitter moods, she shared his interests (both social and professional), and was feelings of horror and despair with which I daily read of the ghastly tragedy which is now being enacted in the fight of the women for political freedom . . . Women like Mrs. Pankhurst . . . are being done to death in a manner which would entail the severest reprisals on the part of the community if practised on any cat or dog in the country.

He related the suffrage question back to his science, explaining that now,

. . . physical force is no longer the sole attribute of living beings. The force which drives the world at its present rate is a purely inanimate commodity, bought and sold in quantities which make the relative differences of natural physical force between warring parties, whether the two sexes or two nations, of no account and therefore no criteria at all nowadays in the arbitrament of human quarrels.

He gave a public lecture at the Caxton Hall, in which he repeated these views at greater length and with more emotion.

Repression, calumny and brutality have been tried long enough. They have formed that political consciousness into a flame which has girdled the earth and welded, for the first time in the history of nations, women as far asunder as the poles in every other respect into one sister-hood remanding justice as human beings and rights as citizens in professedly civilised states . . .

He reiterated his point that science had destroyed any question of the importance of the distribution of physical strength, and then suggested that 'the argument that women on the whole are less divinely endowed with intellect and general capacity than men is too silly to be advanced any longer, I believe, even on anti-suffrage platforms.' His final, conclusive argument was that the twentieth century depends on power, it will, in time, depend on atomic power, and that this power was discovered by the greatest contemporary scientist '. . . *a woman*—No person in the sense of our poor law when it is a question of a privilege granted by the state.' However, this was only possible because she was not English. Had Marie Curie been born in England, she 'would have been classed with paupers and lunatics'. After all this, his conclusion is almost an anticlimax. 'I regard women's cause as won in the fair field of argument and reason. The actual hour alone of victory remains to be fought for.'[62]

As well as these activities, he attended suffrage meetings and from 1913 employed a woman as his personal assistant and deputy and even published a joint article with his wife.[63] Despite this, he never seems to have joined any of the suffrage organisations. He explained to Mrs Arncliffe-Sennett, the founder of the Northern Men's Organisation for Women's Suffrage in Glasgow, that he would not join her organisation because

But it was not only Wells. Soddy's *Matter and Energy*, published in 1912, written at the time of or just after the miners' strike of 1911, shows how his anxieties about the power and use of science were beginning to be grounded in real social and political problems—the 'bitter industrial strife between the wage-earner and the employer' which results from a diminution in the purchasing power of money and the concomitant rise in the price of commodities.[54] This book increased Soddy's popularity in Russia when it was translated in at least two editions in 1913, and where it was described as 'a philosophical generalisation of the concepts concerning the structure of matter and energy that is being produced by the latest trends in physical science'.[55]

Political discussion challenged and changed more than his concerns with the implications of science. As he was being pushed further towards socially critical views by his evaluation of the function of his work, he was also forced to evaluate his personal politics. In discussion in the Beilby household, one particularly frequent subject was women's suffrage. Emma Beilby had been involved with questions of women's political and educational equality for many years. She had, while living in Edinburgh, been one of the first patients and a firm supporter of Sophia Jex-Blake, the 'lady doctor' who had set up a clinic for women and children at her home in Bruntsfield Lodge when she was denied permission to practice in Britain.[56] She retained an interest in similar matters after her move to Glasgow, being asked to become President of a branch of the Infant Health Visitor Association in 1914,[57] and, as we have seen above, Margaret Todd, who provided the term 'isotopes', was a regular visitor to the house.

Emma Beilby shared these interests with her daughter, Winifred, who had presented a basket of flowers to Jex-Blake on her retirement.[58] Despite the family's support of 'female emancipation', Winifred had had a conventional late-nineteenth-century middle-class upbringing. She had never been to school but had been taught at home by governesses and then, showing an interest in art, had studied in Paris for a year or so before working in the Glasgow studios for some time. However, in Glasgow she was also 'an ardent worker for the suffragettes',[59] becoming Vice Chairman [sic] of the liberal Glasgow Society for Women's Suffrage, later a part of the Scottish Federation of the National Union of Women's Suffrage Societies. Her mother, as Lady Beilby, became Honorary Vice-President.[60]

Through this influence, Soddy became interested in the whole of the 'Woman Question'. He wrote both to *The Times* and the *Glasgow Herald* in 1913 in support of the suffragettes and to express his[61]

Soddy looked at these questions in the final chapter of *The Interpretation of Radium* which, especially in the first edition, contains a whole range of speculations going far beyond an explanation of the scientific facts. He began this discussion by describing the benefits which might accrue to mankind through these discoveries: the abolition of poverty through cheap power, the cleansing of the atmosphere through the end of coal burning, the foundation of a new Garden of Eden. In this, it is difficult not to see echoes of *News From Nowhere*, where there were 'no marks of the grimy sootiness . . . on every London building', and a new mysterious form of power, 'the force' of barges on the river, has replaced steam.[50]

Soddy suggested that there might also be problems. Just as Morris's world was the resurrection of an earlier one which had been destroyed by social revolution, Soddy's Eden followed disaster. From the new calculations about the age of the earth, calculations which had sprung from the discovery of naturally occurring radioactivity in rocks, Soddy went on to suggest that since the earth was so much older than had previously been believed, almost any explanation about its history was possible, including a cyclic progression. To illustrate this Soddy returned to the symbol of Ouroboros, the tail-devouring serpent, and suggested that this and other ancient myths give 'some justification for the belief that some forgotten race of men attained not only to the knowledge we have so recently won, but also to the power that is not yet ours?'[51] He then went even further and used these new interpretations to explain the Fall of Man. He advanced the notion of an earlier world where energy had been produced from transmutation which made possible the transformation of deserts, thawing of the ice at the poles and 'mak[ing] the whole world a smiling Garden of Eden.'[52] However, 'one can see also that such dominance may well have been short-lived. By a single mistake, the relative positions of Nature and man as servant and master would, as now, become reversed, but with infinitely more disastrous consequences.' This kind of re-working of the New Atlantis myth was far from new. Writers from Francis Bacon to William Morris had postulated that scientific discoveries could lead both to Utopia and its destruction. What was new here was that Soddy believed he had discovered a particular and real source of power which could be harnessed to provide energy for this new world.

The attractions of this picture must have been enormous. Through the use of science, those who made their living in mines or steel works could be promised a life of ease and plenty. Working class slums could be as clean and comfortable as the bourgeois homes on the hills. Best of all, there would be no more anxiety about money, food and clothing. All could be provided at the touch of a button. Not surprisingly, H. G. Wells picked up and publicised the implications of these ideas in *The World Set Free*.[53]

and the technology on which the contemporary society depended would be useless. The issue of the third edition of *The Coal Question* by Jevons in 1907, in which he suggested importing coal from Australia and Canada, brought home the urgency of the question.

Although Jevons's figures can be questioned, the miner's strike of 1911–12 which 'paralysed the British coal industry'[46] showed what the world would be like if coal stocks should fail before an alternative had been found. As Soddy pointed out, 'at the moment of writing, employers and employed in that [coal] industry are calmly contemplating a complete stoppage while they fight out their differences, and the public experiences a temporary return to the original conditions of life of primitive man.'[47]

However, for Soddy, the problem was more than just a temporary difficulty. Using the Second Law of Thermodynamics, he pointed out that life on this planet depends on the flow of energy 'from its primary atomic reservoirs to the sea of waste heat of uniform temperature.' He also pointed out that the pace of early twentieth-century life had accelerated to the point where the stores of energy in coal, gas and oil and even possible chemical sources were increasingly being used to augment the constant radiant energy of the sun and the 'white fuel' of hydroelectricity. Energy was needed for industrial processes, for agriculture in the production of fertilisers, for domestic uses, and as stores were used up, those processes would inevitably fail, unless an alternative source of energy could be harnessed. 'It looks, therefore, as if our successors would witness an interesting race, between the progress of science on the one hand and the depletion of natural resources upon the other . . . The fact remains that, if the supply of energy failed, modern civilisation would come to an end as abruptly as does the music of an organ deprived of wind.' The only remaining source of energy was in 'the primary stores of atomic energy on the one hand and the waste heat energy of uniform temperature on the other.'[48] Using the latter was impossible, since it went against the apparently immutable Second Law of Thermodynamics which stated that heat does not move spontaneously from a colder object to a warmer one, so the former was the most likely and here Soddy's ten-year fascination with estimates of the amount of energy contained within radium atoms and dreams of harnessing it had a severely practical application. Nevertheless, all the efforts of science to influence the radioactive processes so far[49]

'appear rather like trying to influence the course of a bullet by blowing at it . . . Nothing that is known will affect the transmutation of one element into another . . . It is a mistake to suppose that it is only a matter of time before science succeeds in this quest . . . the release of the energy associated with the structure of the atoms for practical purposes has not yet been brought appreciably nearer by the discovery of radioactivity.'

one of 'the intellectuals', embarking on an autodidactic programme of reading before leaving the city for London.[40] Soddy was ten years older and his bourgeois world of the university probably kept him from this world. There is no direct evidence, at this stage, of his being involved with the ILP, attending meetings or giving lectures for them. However, the general tone of his writings, his concentration on the eradication of poverty and the improvement of society through practical means together with a lack of any economic theories about how to go about it, suggest contact with the kind of socialism which was particular to Glasgow at this period. Glasgow was seen as a centre of 'municipal socialism', but it was also the socialism of Morris' Socialist League. The 'heroes were Jesus, Shelley, Mazzini, Whitman, Ruskin, Carlyle, Morris. The economists took second place. The crusade was to dethrone Mammon and to restore spirit, and to insist that the welfare of the community should take precedence over the enrichment of a handful.'[41] There was, as another historian has described it, 'a style and a fermenting energy which drew the young and uncommitted like bright lights. The sense that a man should act to determine his own fate, that the future was in his own hands and immediately improvable, linked the early socialists to the older radical tradition at its most ardent.'[42] Although the society must have been inviting, there is no evidence of Soddy's involvement in any political party at this stage and he gave no directly political lectures in Glasgow. None the less, his later behaviour and intellectual development seems to stem directly from this background, and therefore it seems likely that he was influenced by it.

However, for the time being, his interests lay in other directions, which were to lead to the formulation of a different vision of the world freed from labour and divisions. In 1906 he wrote to *The Times* to support the theory which he had formulated with Rutherford that radioactivity provided a source of enormous quantities of heat.[43] He concentrated increasingly on one area, that of the use which could be made of the energy contained within atoms if only it could be harnessed and produced as required. This clearly shows Beilby's influence. Beilby had been concerned with the economical use of fuel for a number of years and had used his presidential address to the Society of the Chemical Industry in 1899 to review the subject. He had subsequently been appointed as the chairman and director of the Fuel Research Board.[44] One of the concerns of this board was to find a replacement fuel for coal and, throughout his time in Glasgow, Soddy was also working with Beilby on a number of schemes connected with the economical production of fuel. They attempted to design new batteries, and Soddy patented some of the more promising attempts.[45] The work was important since predictions by Jevons and others from the 1890s onwards suggested that stocks would run low

Technical College where he personified the intention of the Board to enlist 'the sympathy and co-operation of representative members of the leading industries of the district.'[37] From both these positions, Beilby was able to show his commitment to education. He also, in his home, showed himself to be liberal (with a small 'l') and concerned with a range of political and social issues.

Until his move to Glasgow, there is no sign that Soddy was involved at all in politics. His reading, his social life and his work had all centred on science. He was perhaps beginning to demonstrate an appreciation of the natural landscape, especially hills and wilderness, sparked by his visit to the Lake District with Harold Carpenter and then his trips to Australia and the North American continent, but his concern for the inhabitants and cultures of those worlds resembled that of the tourist. In contrast, the Beilby family had a range of political and cultural concerns normal for the early twentieth-century bourgeoisie which fascinated the young Englishman. His experience of a Chapel-centred childhood was very different from the open house and open minds policy practised by the Scottish middle class household that he now experienced.

When George Beilby and his family moved to Glasgow from Edinburgh they had moved into University Gardens, an imposing terrace of Edwardian houses overlooking the University and occupied by many of the professors and others connected with the institution. Within this terrace there operated an informal but highly influential organisation on behalf of the university.[38] Almost anyone involved in higher education in the city either lived in the terrace or had connections there. As Soddy and many others were to find, it also provided a social centre for new staff and researchers where they could avoid some of the loneliness associated with new jobs and towns. Further, George and Emma Beilby's son and daughter lived with them and they were sociable and amiable hosts; their house was a centre for young people from many areas of the city as well as for the rather older university circle. At the regular Sunday afternoon socials which were held at 11, University Terrace, conversation would range broadly outside the immediate concerns of the university, to cover national political and social questions.

Glasgow was a peculiarly appropriate place for such discussions as it was one of the centres for the growth of socialism and particulary the ILP in Britain. It offered the younger generation 'a rich variety of red-blooded socialist things to do', beginning with a socialist Sunday school, then joining the Clarion scouts to distribute ILP literature and subscribing to *Forward*.[39] A description of this is given by the poet Edwin Muir, later, like Soddy, to be involved with the New Europe Group. As a dissatisfied clerk he joined the Clarion scouts, attended innumerable Sunday evening lectures at the Metropole Theatre and then became theories of evolution on the grounds that the earth was not old enough for such changes to take place. Huxley, amongst others, had argued against him and their debate had become loudly public. Although Huxley and his supporters had evidence to support their case, they could find none to refute Kelvin's argument. The question was brought up again at the British Association meeting in 1894, but again no evidence could be produced to shift Kelvin. Then, in 1904, Rutherford had challenged Kelvin, arguing that 'the discovery of the radio-elements, which in their disintegration liberate enormous amounts of energy, thus increases the possible limits of the duration of life on this planet, and allows the time claimed by the geologists and the biologist for the process of evolution'.[34] Finally, Kelvin was persuaded and at the British Association meeting that year 'he acknowledged that the phenomena were atomic'.[35] As a result of this, when Soddy arrived in Glasgow, Lord Kelvin, now Chancellor of the University of Glasgow, seriously discussed with his wife whether it would be desirable to invite the young scientific revolutionary to Largs. They did so, and although areas of scientific disagreement remained, Kelvin and Soddy found their mutual respect had increased.[36] This was just as well, as will be discussed in more detail below, since the Glasgow academic community was close-knit, centring on University Terrace where both the Kelvins and the Beilbys had houses, so disapproval from any one member would lead to complications in the social life of the university.

Soddy also continued to address a popular audience. He had written articles in *The Times, the Contemporary Review*, and other periodicals produced for a popular audience. He gave public lectures to the Glasgow electrical engineers in 1906 and then a series of 'six experimental' public university lectures in 1908. He had published the latter as *The Interpretation of Radium*, and they were advertised as 'an account, with experiments', of the researches into radioactivity to date. These experiments were the ones which had proved so popular in Australia. In all these works, Soddy continued to suggest the possible practical applications of radioactivity.

Soddy's non-academic work stemmed from a fortunate convergence of two forces in his life. He had found, in Australia, that he enjoyed and was successful lecturing to a popular audience. The popularity of his written work in journals and books for a general readership showed that he had considerable abilities in this field. He was enjoying the opportunity to move out of pure science to see some of its applications in industry. Beyond all these factors, contact with George Beilby and his circle brought home to Soddy the political ramifications of his work in different ways. The first was in the field of further education. As well as his 'unofficial' position with regard to the university, Beilby was Chairman of the Governors of the Glasgow and West of Scotland

Mme Curie for the radium contained in the International Standard, as a personal tribute to Mme Curie and her work.'[29] However, in general, the supply remained inadequate. Precious samples were lent and borrowed, complicated arrangements were made between laboratories to provide some part of the radium emanation for a particular purpose.[30] In addition, theories about potential uses, medical and industrial, for radium were increasing but could not be tried without quantities of raw material. This demand could be satisfied only by discovering a new source of radium, by finding a means of hastening the natural processes of its production or by finding an alternative radioactive substance. Of these, the latter seemed the most promising. The key figure here was Otto Hahn.

Hahn had worked with Ramsay in 1904–5 after Soddy's departure and, while there, had discovered the new element radio-thorium which was to become a useful alternative to radium. After he returned to Germany, Hahn discovered mesothorium, another radioactive product of thorium. There was a rumour, which Soddy believed, that Hahn was keeping some of his discoveries secret for commercial purposes. This was not totally unlikely. The profit from such a process could have been large since thorium, from which the new elements had been separated, was fairly common, and already in use in the manufacture of gas mantles. Soddy's response was himself to work to discover the process of separation of mesothorium from monazite sands, and he registers preliminary specifications for a patent on mesothorium production in 1911 to which he then allowed free access.[31] He had both a nationalist and a scientific motive for this, each of which was to have marked long-term effects.

Monazite sands were found in limited areas, including particular caves in Brazil. Rutherford discovered that the Brazilian government was supplying the Germans with unlimited quantities of the sands free of charge.[32] When he heard of this, Soddy believed that there was a potential danger in Germany having unlimited supplies of such a material and began a long term and ultimately unsuccessful campaign for the import of similar sands from Travancore in India for British use, which extended throughout his time at Oxford and even led him to take a trip to India to see the possibilities for himself.[33] The scientific motivation was his absolute opposition to any attempt to benefit financially from scientific discoveries as he had already shown by his reaction to doubt Ramsay's behaviour when it seemed possible that he was attempting to gain financially from his work.

While at Glasgow, Soddy had demonstrated his abilities as a chemist through his research, especially the discovery of isotopes and his publication of the periodic table, and persuaded almost everyone by his proofs. However, there was one notable exception. Lord Kelvin refused to be convinced. He had been arguing since the 1860s against Darwin's violation of scientific etiquette'.[26] A more recent account of this episode points out that although Fajans and Soddy came to the same findings, their conclusions were entirely opposed. Fajans believed that he had shown that his results were evidence against the nuclear origin of the changes involved while Soddy saw that he had evidence for 'a complete and correct version of the Periodic Table displacements associated with radioactive changes.'[27]

In the light of what has been said about his character and lack of diplomatic skills, it comes as no surprise to learn that Soddy took part in a number of other disputes during his time at Glasgow. He and Fleck were involved in another aspect of the controversy surrounding the chemically inseparable elements. This was connected with Antonoff, then an unknown Russian research student working with Rutherford and afterwards in Russia. The problem arose over Antonoff's claim to have isolated uranium-Y, a new radio-element. Fleck and Soddy were unable to repeat the experiment and so doubted the accuracy of the findings. This resulted in an acrimonious exchange between Fleck and Antonoff before the British scientists were finally convinced by the publication of further details of the experiments.[28] Unimportant in themselves, these episodes show the extent to which Soddy's character predisposed him to insist on the correctness of his own position in arguments—a disposition which had already led to initial difficulty with Rutherford and was to lead to far more awkward situations in the future. At this stage he was arguing about 'facts' which could ultimately be 'proved'. Over the next decades he was to argue over matters of belief and impression and in these conflicts he was not always able to carry his case. As a result his reputation for unreasonable behaviour grew.

Another controversy demonstrates Soddy's inflexible insistence on what he regarded as correct moral behaviour, and his motives seem to have been unimpeachable. The supply of radium for research was a constant obstacle which meant that experiments had to be designed in ways which took account of this limitation or were even postponed until sufficient supplies were available. Since 1903 Soddy had been active in attempting to increase the supply, both for research and for medical applications. He found a temporary and personal solution through the generosity of the Beilbys. Through his contact with Soddy, George Beilby became interested in the problems associated with radium and in 1912 he and Mrs Beilby were able to offer practical assistance. One of the myriad problems associated with radium was the lack of an internationally accepted standard of measurement for radiation. Work on this standard took place through 1911 but the sample used belonged to the Curies. In April of 1912 Soddy was able to write to Rutherford, 'Dr and Mrs Beilby have generously donated the sum necessary to refund subject of the radio-elements, their possible transformation or transmutation and their relationship with the conventional periodic table of the elements was one which exercised the radiochemists throughout the scientific world after the initial discoveries of Rutherford and Soddy. Increasing numbers of examples of chemically inseparable elements were discovered, leading to more problems in assigning places in the periodic table. By 1910, Soddy was convinced that 'cases of chemical inseparability were actually chemical *identities*' and was the first to 'declare that these were chemically identical elements'.[22] This was the statement of the discovery for which he received the Nobel Prize in 1921. The actual term 'isotopes' came later from a conversation at the Beilby's house with Dr Margaret Todd, a distinguished physician, writer and friend and biographer of Sophia Jex-Blake, who offered the word as the Greek for the phenomenon he described. However, the remaining difficulty was how the radioactive elements changed in the course of radioactive emission and then how they fitted into the periodic table. These questions had been the concern of chemists since the nineteenth century but were now given added urgency by the discovery of each new element. The periodic table could accommodate only one element in each box, but refinements were offered like that of Crookes in 1886 which attempted to theorise elements evolving from one to another. None was successful, largely because the necessary discoveries were still in the future. In 1911, Soddy offered a partial solution when he published *The Chemistry of the Radio-elements*, in which he proposed that after the emission of the alpha particle, the daughter element occupied a place two boxes away in the table. Like his other works considered earlier, this was translated into Russian. It was published in St Petersburg in 1913 with a preface by V. A. Borodovsky, one of the first Russian radiologists, who said that it 'represents an exceptional phenomenon in the literature on radioactive substances . . .' With this, Soddy's work continued to increase in popularity in Russia.[23]

Too little was known about the results of beta emitters for theorising at this stage, but Soddy set Fleck on to experiments into the problem in 1912. By the end of 1912 Soddy, Fajans, Hevesy and Russell were all nearing a solution to the problem. Hevesy and Russell both produced partial solutions and then Fajans and Soddy, from different directions, both produced essentially correct versions early in 1913.[24] For some reason, although Fajans's was published two weeks earlier, Soddy did not acknowledge it in his final version. The probable explanation for this is that he was so closely involved in the work of his laboratory that he did not realise that he was applying what he had read to a current problem: 'It is likely that Fajans' paper caused everything to fall into place in Soddy's mind and Soddy, being so close to the material, could not later sort out what he had borrowed from another.'[25] However, it was 'a surprising

'Cavendish scientists'.[17] This was the beginning of something Soddy felt for the rest of his life—that there was some kind of conspiracy or 'claque' amongst the Cambridge scientists which always worked to his disadvantage. It is somewhat ironic that this was recognised by a later generation of Cambridge men who blamed it on the arrival of Rutherford at the Cavendish and his desire for a team of 'his boys'.[18] At first, Soddy also found it difficult to adjust to Scotland and Scottish ways. In one of the letters to Rutherford he complained that[19]

Scotch Universities are like their Church very much a part and organisation of their national life, much less free and independent much more a laboured and elaborate creation than a spontaneous and natural intellectual movement . . . I feel rather like a loose cog that has been pitched into a wonderful piece of mechanism . . .

Here, in this remarkably apt simile, Soddy describes what was to become a perennial problem, and not one peculiar to Scotland; he was always to be a loose cog in any institution. He was never able to make the necessary politic adjustments to his views, or even the airing of them, which facilitate the smooth running of any large organisation, but insisted on saying precisely what he thought and saying it loudly and publicly. Unlike Rutherford, Soddy was never capable of dealing with committees and groups and so eventually they in turn would disregard him.

On occasion, there was more than just a general sense of dissatisfaction. For example, soon after the letter quoted above, Soddy complained to Rutherford about his conflict with the university authorities and the other science faculty over a relatively trivial matter. Soddy had applied for a pay rise for his mechanic but after six months no answer had been given so the man had left. He had still not been replaced some four months later and, as a result, Soddy argued that he had been unable to continue his research. He had decided against asking the authorities for anything else because he felt that this was part of a considered attack on him by Professor Ferguson, who had held the Chair since 1874, and other members of the scientific faculty of the university. Soddy believed that there was a plot to reduce his independence and to persuade him to take a subordinate position within the department instead of his independent lectureship.[20] No further details of this matter are now available, but there seems to have been some exaggeration of the situation by Soddy since he declined Rutherford's offer of facilities for research in his laboratories in Manchester. Even 'the opportunity of working in your inspiring company again' did not make it seem worth moving complicated and difficult apparatus.[21]

Despite his anxieties about work and interdepartmental relationships, Soddy's research continued successfully during this period. This whole was particularly proud of another of his Glasgow students, Alexander Fleck. He was the archetypal working lad made good—had been to the Glasgow Technical College, then started work as Soddy's laboratory boy, gaining his research degree before moving into industry and finally becoming chairman of ICI. With these facilities, time, space and people, available, Soddy was initially able to accomplish research projects around what were then seen as the remaining problems of radioactivity, especially the chemically inseparable elements.

However, his delight with the circumstances of his arrival in Glasgow in the autumn of 1904 was short-lived. By the following May he was complaining in one of his regular letters to Rutherford about the unfinished state of his laboratories, and by the autumn he attacked the architecture of new laboratories in Britain.[12] What might have been the real cause of his dissatisfaction was revealed in his letters in the spring of 1906. Much of his enthusiasm for his Glasgow post had centred on the opportunities for research, but by this time, eighteen months after his appointment, he found that his teaching and administrative load had increased to the point where he had given more than a hundred lectures over the winter session as well as practical classes so that 'research was out of the question'.[13]

This is clearly an exaggeration. With his students Soddy had embarked on an ambitious programme of research, but this was time-consuming, and so he could ill afford the demands of lecturing to undergraduates. Each of his lectures was written by hand in a series of blue hardback notebooks, with the text of the lecture on the right hand page and the other left blank for experiments, questions or other interesting points. Once written, they could be given over and over again with relatively straightforward updating but the effort of producing them at the rate of two a week must have been great in addition to of his other tasks, taking time from research which he could ill afford. Soddy's ambition was, as he explained in October 1906, a Chair of Chemistry, ideally in England.[14] For this, original discoveries reported in the professional journals were needed. By this stage in his career he had done well, but not spectacularly so. 'So far, Soddy's name had been linked with those of senior and already famous investigators,'[15] Rutherford and then Ramsay. It was now time to show that he could also work on his own.[16]

Soddy's anxieties were increased by Rutherford's appointment as Professor of Chemistry in Manchester and were demonstrated in his letters, where increased sarcasm and dissatisfaction revealed his jealousy at his friend's success. England rather than Scotland beckoned as the centre of the scientific world and Soddy was soon attempting to move south. Unfortunately, there were advertisements for professors of physics but nothing for chemists, and anyway posts were always awarded to

Soddy is recorded, but it seems likely that it took place because when the new post was decided upon, Soddy was remembered.

He seemed ideal. The University of Glasgow was looking for someone 'whose position in the Scientific world is already established' and were 'fortunate enough to ascertain that Mr. Frederick Soddy, one of the most brilliant workers on the subject of Radio-Activity would be willing to offer himself as a candidate'. Members of the appointment board contacted Ramsay who confirmed Soddy's brilliance. They proposed to offer the post without interview and to delay further plans for the improvements to the chemistry laboratory until his arrival so that he could make his views known.[7]

At this stage of his career, this post suited Soddy very well. He was appointed primarily to establish a research school and so seems, initially at least, to have been required to do little lecturing. He therefore avoided the difficulties of teaching rowdy Scottish undergraduates who considered their role to be to test the nerve of the lecturer by riotous behaviour. As a contemporary wrote, they were 'difficult to handle, self centred and self assertive with no social sense'.[8] We have seen how Soddy reacted to one kind of boisterous behaviour when he first met Rutherford, and there is no indication that he would have fared better here, especially since he remained shy and aloof, with a voice which was 'high pitched and very southern English in accent', as one of his early students was to remember.[9] More directly, he needed all his energies to organise facilities for research. Throughout his scientific career Soddy was to argue that research departments should be free from the ties of undergraduate teaching. He asked that 'the creative aspect of science be considered and provided for totally independently of the professional and educational aspects.'[10] Immediately, he was in this relatively privileged situation. He was able to organise a research programme around four main topics connected with radioactivity in the new laboratories known as 'the Mud Huts' which had been built for the advanced courses in organic, inorganic and metallurgical chemistry.

He then acquired a group of ten enthusiastic research students who were attracted by his growing reputation with whom he got on well. These included T. D. MacKenzie and Miss A. F. R. Hitchins, who worked with him for a number of years on the establishment of the disintegration theory by proving the growth of radium from uranium using the 50 kilograms of uranyl nitrate given by George Beilby for the purpose. Others were A. J. Berry who worked on rarefied gases and A. S. Russell who worked firstly on γ-rays and later contributed to the work on isotopes. Miss Ruth Pirret studied uranium radium ratios; A. J. Cranston's work was on the parent of actinium and H. Hyman worked on the atomic weight of lead from thorium minerals.[11] Soddy

CHAPTER 3
Glasgow

Soddy arrived in Glasgow in the autumn of 1904 at the beginning of what was to prove, in his own words, his 'most productive period'.[1] Not only was his professional career to prove that he was successful as an individual, not just as the junior partner of Rutherford or Ramsay, but outside science he was to be introduced to a range of concerns and interests which were to occupy him for most of his life. Most importantly, he was to meet his future wife. Each of these aspects was brought about through a chance meeting in 1902 with George Beilby, a Scottish industrialist with a wide circle of acquaintances in educational as well as industrial areas.[2] The first and fundamental assistance rendered by Beilby was professional. Beilby seems to have used his influence in Soddy's favour when the University of Glasgow, having embarked on a programme of modernising their chemistry department in the summer of 1903, had decided to appoint an Independent Lecturer in Physical Chemistry around whom a research school could be started.[3]

George (later Sir George) Beilby acted in an advisory capacity in this matter. He had no official standing in the university but was keenly interested in the teaching of scientific and technical subjects, becoming chairman of the governing body of the Royal Technical College in Glasgow in 1907. As a result he knew many members of the department and had acted as unofficial advisor to the science departments before.[4] Professionally, Beilby had a number of interests which qualified him to give advice. He had spent his working life in the chemical industry and at this time was a director of the Cassel Cyanide Company. He was president of the Society of Chemical Industry and also had a long term concern with the economical uses of fuel which led to his membership of a number of government commissions on the subject.[5] He was a regular attender at meetings of various professional associations and as a result had a wide range of acquaintances within the industry. Amongst these was Soddy's schoolfriend Harold Cort Carpenter who had been appointed head of the metallurgical department of the National Physical Laboratory. Carpenter wanted to discuss his new post and their common interest in metallurgy with Beilby, which they did at the Glasgow meeting of the British Association in 1902.[6] Coincidentally, Soddy had returned from Montreal in order to attend this meeting, and at the same time to share a holiday in Scotland with Carpenter. No meeting between Beilby and

54. Rutherford Papers, Cambridge University Library, AD 7653, S109; Eve, A. S. (1939). *Rutherford: Being the Life and Letters of the Rt. Hon. Lord Rutherford, O.M.*, p. 99. Cambridge University Press.
55. Soddy, F. (1909). *The Interpretation of Radium*. John Murray.
56. Rutherford Papers, Cambridge University Library, AD 7653, S117.
57. Rutherford Papers, Cambridge University Library, AD 7653, S121 and S122.
58. For what follows see the article by Krivomazov in Kauffman, G. B. (ed.) (1986). *Frederick Soddy (1877–1956): Early Pioneer in Radiochemistry*, pp. 115–40. Reidel, Dordrecht.
59. Quoted in the article by Krivomazov in Kauffman, G. B. (ed.) (1986). *Frederick Soddy (1877–1956): Early Pioneer in Radiochemistry*, p. 119. Reidel, Dordrecht.
60. Rutherford Papers, Cambridge University Library, AD 7653, S117.
61. Rutherford Papers, Cambridge University Library, AD 7653, S120.
62. Rutherford Papers, Cambridge University Library, AD 7653, S121.
63. Rutherford Papers, Cambridge University Library, AD 7653, S122.
64. Soddy, F. (1904). The Wilde Lecture VIII: The Evolution of Matter as Revealed by the Radioactive Elements. Given on 16 March 1904. *Memoirs and Proceedings of the Manchester Literary and Philosophic Society*, **48**.
65. Jenkins, J. G. (1985). Frederick Soddy's 1904 Visit to Australia and the Subsequent Soddy–Bragg Correspondence: Isolation from Without and Within. *Historical Records of Australian Science*, **6**, 56.
66. Rutherford Papers, Cambridge University Library, AD 7653, S123.
67. Badash, L. (ed.) (1969). *Rutherford and Boltwood: Letters on Radioactivity*, pp. 57 and 79. Yale University Press, New Haven.
68. *Memoirs*, pp. 102–3.
69. Jenkins, J. G. (1985). Frederick Soddy's 1904 Visit to Australia and the Subsequent Soddy–Bragg Correspondence: Isolation from Without and Within. *Historical Records of Australian Science*, **6**, *passim*.
70. Quoted in Jenkins, J. G. (1985). Frederick Soddy's 1904 Visit to Australia and the Subsequent Soddy–Bragg Correspondence: Isolation from Without and Within. *Historical Records of Australian Science*, **6**, 159.
71. Jenkins, J. G. (1985). Frederick Soddy's 1904 Visit to Australia and the Subsequent Soddy–Bragg Correspondence: Isolation from Without and Within. *Historical Records of Australian Science*, **6**, 159.
72. *The Morning Herald* (Perth), 25 July 1904, p. 6.
73. *Memoirs*, p. 111.
74. *Memoirs*, pp. 111–12.
75. *Memoirs*, p. 112.
76. *Memoirs*, p. 113.
77. Rutherford Papers, Cambridge University Library, AD 7653, S127.
78. Rutherford Papers, Cambridge University Library, AD 7653, S127.
79. *Pioneer*, p. 142.

25. Wilson, D. (1983). *Rutherford: Simple Genius*, p. 152. Hodder and Stoughton.
26. Wilson, D. (1983). *Rutherford: Simple Genius*, p. 147. Hodder and Stoughton.
27. Quoted in Wilson, D. (1983). *Rutherford: Simple Genius*, p. 147. Hodder and Stoughton.
28. Letter from Soddy to Rutherford in the Rutherford Papers, Cambridge University Library, AD 7653, S94.
29. *Pioneer*, p. 66.
30. Eve, A. S. (1939). *Rutherford: Being the Life and Letters of the Rt. Hon. Lord Rutherford, O.M.*, pp. 77–8. Cambridge University Press.
31. For a discussion of McGill at this point see the articles brought together in *Rutherford and Physics at the Turn of the Century*, (ed. M. Bunge and W. R. Shea). (1979). Dawson, New York.
32. Quoted in Selove, R. E. (1989). From Alchemy to Atomic War: Frederick Soddy's 'Technological Assessment' of Atomic Energy, 1900–1915. *Science, Technology and Human Values*, **14**, No. 2, 167.
33. See Chapter 3, p. 56.
34. For a full discussion of this question, see Selove, R. E. (1989). From Alchemy to Atomic War: Frederick Soddy's 'Technological Assessment' of Atomic Energy, 1900–1915. *Science, Technology and Human Values*, **14**, No. 2, 163–94.
35. Trenn, T. J. 1986). Frederick Soddy. Introductory chapter in *Frederick Soddy (1877–1956): Early Pioneer in Radiochemistry*, (ed. G.B. Kauffman). Reidel, Dordrecht.
36. Quoted in *Pioneer*, pp. 83–4.
37. These are reproduced in *Collected Papers of Lord Rutherford of Nelson*, Vol. 1, (ed. Sir James Chadwick). (1962). Allen and Unwin.
38. Feather, N. (1977). Foreword to Trenn, T. J. *The Self-Splitting Atom: The History of the Rutherford-Soddy Collaboration*. Taylor and Francis Ltd.
39. Wilson, D. (1983). *Rutherford: Simple Genius*, p. 148. Hodder and Stoughton.
40. *Pioneer*, p. 82.
41. Soddy, F. (1903). Some Recent Advances in Radioactivity. *Contemporary Review, May 1903*, 719.
42. Soddy, F. (1903). Some Recent advances in Radioactivity. *Contemporary Review*, May 1903, 719
43. Rutherford Papers, Cambridge University Library, AD 7653, S94. An account of the Ramsay/Soddy experiments can be found in Travers, M. W. (1956). *A life of Sir William Ramsay K.C.B., F.R.S.*, Chapter 14. Edward Arnold.
44. *Memoirs*, pp. 78–9.
45. *Memoirs*, pp. 78–9.
46. Quoted in Travers, M. W. (1956). *A life of Sir William Ramsay K.C.B., F.R.S.*, p. 214. Edward Arnold.
47. Rutherford Papers, Cambridge University Library, AD 7653, S95.
48. Rutherford Papers, Cambridge University Library, AD 7653, S117 and S125; Badash, L. (ed.) (1969). *Rutherford and Boltwood: Letters on Radioactivity*, p. 57. Yale University Press, New Haven.
49. Rutherford Papers, Cambridge University Library, AD 7653, S95.
50. *Pioneer*, p. 104.
51. *Memoirs*, p. 89.
52. Rutherford Papers, Cambridge University Library, AD 7653, S105.
53. Rutherford Papers, Cambridge University Library, AD 7653, S108.

future his name was forgotten in that context, at the time if anything he was popularly the better known of the collaborators. Further, his later work with Ramsay laid the foundations for his later discovery of isotopes. Nor was his success only professional. Although he remained personally difficult and shy he discovered a real talent for popular writing and lecturing which was never to leave him.

Notes and references

1. *Pioneer*, p. 55.
2. Soddy, F. (1933). Reminiscences of McGill, 1900–1902. *Old McGill* Annual, **1936**, 17.
3. Cruikshank, A. D. (1979). Soddy at Oxford. *The British Journal for the History of Science*, **12**, 278.
4. Williams, T. I. (1964). Frederick Soddy and the Concept of Isotopes. *Endeavour*, **23**, 54.
5. *Pioneer*, p. 55; *Memoirs*, p. 41.
6. Wilson, D. (1983). *Rutherford: Simple Genius*, p. 166. Hodder and Stoughton.
7. *Memoirs*, p. 42.
8. *Memoirs*, p. 66.
9. *Memoirs*, p. 67.
10. Interview between the author and Marjorie Soddy, July 1990.
11. *Memoirs*, p. 73.
12. *Memoirs*, p. 69.
13. For example, Soddy's *The Story of Atomic Energy* (1949) begins with an account of the history of science.
14. Professor Frederick Soddy Papers, Bodleian Library, Oxford, MS Eng Misc b179, f 99.
15. Quoted in Selove, R. E. (1989). From Alchemy to Atomic War: Frederick Soddy's 'Technological Assessment' of Atomic Energy, 1900–1915. *Science, Technology and Human Values*, **14**, No. 2, 166.
16. Sinclair, S. B. (1986). Radioactivity and its Nineteenth Century Background. In *Frederick Soddy (1877–1956): Early Pioneer in Radiochemistry*, (ed. G. B. Kauffman), pp. 43–53. Reidel, Dordrecht.
17. Wilson, D. (1983). *Rutherford: Simple Genius*, p. 154. Hodder and Stoughton. See also Jaki, S. L. (1979). The Reality Beneath: The World View of Rutherford. In *Rutherford and Physics at the Turn of the Century*, (ed. M. Bunge and W. R. Shea), pp. 124–39. Dawson, New York.
18. *Pioneer*, pp. 64–5.
19. The text of Soddy's contribution to this debate is found in Professor Frederick Soddy Papers, Bodleian Library, Oxford, MS Eng Misc b179, ff 101–2.
20. Wilson, D. (1983). *Rutherford: Simple Genius*, pp. 148–50. Hodder and Stoughton.
21. *Pioneer*, p. 80.
22. *Pioneer*, p. 80.
23. *Memoirs*, p. 65.
24. See Chapter 1, p. 16.

for his disappointment and he was pleased with the ceremonial greetings of garlands of flowers.

On arrival in San Francisco he discovered that, because he had sent his lecture fees straight home, he could not afford to cross to the Yosemite Valley and the Grand Colorado Canyon as he had intended, and in fact he never did get to see them. Instead, he went to the 1904 St Louis Exposition, where, in the hall of machinery,[76]

I stood and wondered as many others have done, whether the man of human flesh and blood and unique power of mind, was not in danger of being mastered by and becoming the sport of these colossal creations of his handiwork. No doubt each machine had its use for the benefit of mankind, but somehow, these massive inanimate monsters looked sinister and inhuman, particularly as in the centre of the great hall they had put a forty-foot high statue of the god Vulcan in cast iron, no doubt there to guard his infernal machinery.

These were the first intimations of the doubts that were to assail him later about the wisdom of scientific discoveries and the direction in which they would take the human race. During the next twenty years these fears about science were to be increased. As Soddy became conscious of the power of science, and specifically the potential energy within atomic disintegration, he became increasingly disturbed that this was not put to use for the improvement of society. During the First World War these fears worsened as he came to contemplate the full horror of the possibility of atomic weapons. Eventually the image of the blacker side of science which began with his mental image of the statue of Vulcan returned to him in 1945 in the aftermath of the first atomic bomb.

Soddy was 'naturally a little sick of travelling and . . . glad to get home again' after his travels: '. . . it has been a change if not exactly a rest.'[77] One of his first actions after his return was to visit his father in Eastbourne where he related stories about his Australian and world tour to which 'he listened with the eagerness of a child'. Even more importantly,[78]

I wanted to express my gratitude to him for so ungrudgingly assisting me financially during the ten years from my leaving school. I was pleased to tell him that, at long last, I could consider myself financially independent with a salary of £400 a year as lecturer in physical chemistry and radioactivity (the first ever) at Glasgow University . . . that I considered I owed my career in pure science solely to him.

This was good news, but Soddy was to regret that his father died in 1911 before the 'few honours that came my way'.[79]

On his return from Australia, Soddy could look back on what was probably the most scientifically successful period of his life. His co-operation with Rutherford had given him a key role in what some would argue was the single most important scientific research of this century. Although in set in Perth, at a time when a labourer could expect to earn about £2 a week and the rent of a house was ten shillings. Even so, the demand had been underestimated and, at the first lecture on 20 June,[70]

> The reputation of the lecturer, and the subject matter of the lectures ... were evidently responsible for the anxiety which last night's overflowing audience showed ... and hundreds (including many holders of Season Tickets) were unable to obtain admission.

As a result, a larger hall was found and the first lecture repeated. The Australian press reported the lecture approvingly, referring to his 'clear though rapid enunciation' and the use of 'a simple but effective analogy to more clearly illustrate his meaning' and 'the apparatus with which Mr. Soddy conducted a series of striking experiments'.[71] As a result, audiences of 1000 to 1500 were attracted and Soddy's suggestion that opera glasses would help those at the back to see the experiments must have been very useful.

Overall, the whole trip seems to have been a great success and from all the accounts he emerges as a warm, modest and generous man. At the concluding lecture on 23 July he was thanked by the Premier of Western Australia who moved a vote of thanks but pointed out that this was hardly necessary as the attendance and enthusiasm of the people had shown their gratitude. He also felt that the visit had given a renewed impetus to the movement for the establishment of a new university.[72]

From Adelaide, Soddy travelled to New Zealand for a brief visit, but he had little contact with scientists. He landed on the South Island and tried some climbing but it was midwinter so soft snow was a handicap.[73] The high spot of Soddy's time in New Zealand was a visit to the Rotorua Hot Springs district on the North Island where he found compensation for just missing the eruption of a volcano. He was guided by a Maori princess who took him to see the 'soaping' of a geyser:[74]

> Some pounds of yellow soap cut into small pieces were dumped into the orifice and everyone fled to a safe distance save one very old Maori wizard who boldly approached the vent waving a wand and emitting an incantation in the Maori tongue. We watched spellbound, and then quite suddenly, hearing I suppose a warning grumble from below, he picked up his skirts and fled at an incredible speed for such a Methuselah. He was only just out of range when a great fat column of boiling water shot up into the air—speaking from memory—over one hundred feet.

After leaving New Zealand he crossed the Pacific and, like many visitors before and after, Soddy found the islands not so idyllic. 'I remember being sadly disillusioned with one of the smaller Pacific islands at which we called—the heat and the flies! and half the native population suffering from elephantiasis!'[75] However, Honolulu made up

Gossip about Ramsay was increasing to the point where 'pointed references to the ha'penny press' were made to Soddy while he was in Manchester. After Soddy's departure 'such a series of "discoveries" that never stood the test of time issued from his laboratory that the competent workers in radioactivity disbelieved *all* announcements . . .' and by the autumn of 1905 Boltwood was using the term 'Ramsamania' to describe unsatisfactory results.[67]

The time spent with Ramsay had been nothing like as productive as that with Rutherford. But Soddy completed his programme of work in London. He had proved that radon was a noble gas and secondly that helium was produced in the course of the radioactive decay of radium. These were worthwhile discoveries, but there is no sign of the kind of excitement which had been a feature of the collaboration with Rutherford nor of the kind of inspirational relationship which had led to such success in Montreal.

Once liberated from the laboratory, the trials of London and the difficulties associated with working with Ramsay, Soddy was free to enjoy his trip to Australia with his customary gusto. He decided to make the most of his opportunities by travelling round the world and seeing as many countries as possible *en route*, and so booked a passage to travel outwards by Suez and home again via the USA. The ticket for this cost £120 for everything except sleeping berths on the transcontinental railways across America. He arranged for his apparatus, all of which was to be taken with him, to be packed for the journey and he was ready to set off. The voyage out on the *Australia* was uneventful. Aden was a 'God-forsaken place to have to live in' but in contrast 'Ceylon was such an earthly paradise that I vowed, then and there, to try to revisit it.'[68] He did in fact revisit Ceylon in 1937 when its beauty again served to provide some happiness in a dark period of his life.

He arrived in Perth in June 1904 to begin his lectures. The series of six was called 'Radium and Modern Views on Electricity and Matter', and he was to give them in Perth, Fremantle and then the country centres of Kalgoorlie, Coolgardie, Northam, York, Albany and Bunbury, all involving a considerable amount of travelling.[69] In all these lectures, Soddy tried wherever possible to use language suitable for his audience, to find relevant metaphors and examples and to use experiments to illustrate his points. Of the latter, the most spectacular was probably discharging an electric current through several Leyden jars and a human chain before it lit a lamp at the end of the chain. The result was that his lectures grew in popularity as his reputation spread. Even before his arrival, it had been anticipated that the subject matter would prove attractive to many people and so a largish hall had been booked. A relatively high charge of one shilling per lecture or five shillings for the series had been move as soon as I get any opportunity at all. I am in fact debating writing to Dixon at Owens [Manchester] as to whether he can renew his earlier offer . . . I have to get something to do now in the teaching line or I shall get absolutely out of regular work.

Instead, he applied for a fellowship at St John's, Oxford, and a Lectureship in Chemistry in Glasgow University. He had some preferences, as he confided to Rutherford in January 1904, 'If I could secure the double event it would be a stroke of luck, but the latter is rather tempting to me of itself as it would enable me to detach myself from University College and continue along my own lines.'[61] Leaving Ramsay's laboratory was becoming more urgent all the time, and he repeated his concern in his next letter to Rutherford a fortnight later: 'as you know, I am anxious to continue on my own and I hope something will turn up soon that will enable me to do so.'[62] This period of unhappiness continued through February when he noted that he had suffered from a succession of colds and was now fairly certain that the fellowship at St John's would not be given for science.[63] It is typical of the general dispirited air which surrounds his correspondence at this time that the prestigious Wilde Lecture, given on 23 February, is mentioned only in an aside during March.[64]

At about this point an offer was made which provided at least a temporary respite. R. D. Roberts, Registrar of the London Board to Promote the Extension of University Teaching, wrote to offer Soddy 'an opportunity of taking an interesting sea voyage with expenses paid' to Western Australia where he would be expected to give a series of about fifteen lectures.[65] This would probably have tempted Soddy whenever it was offered, but at this nadir of his fortunes it must have seemed the only chance of relief.

The hand of fate must have seemed particularly unkind during the next few weeks. Soddy's acceptance of the Australian trip was announced on 3 March and on 12 March he heard that he had been appointed to the Glasgow post. This was unexpected: 'I was much surprised as the cautious Scot seems for once to have taken a bold move and to have elected me largely on trust.' Accepting this post meant that the Australian trip was 'rather a nuisance' but he could not refuse it. Although it would be 'rather a rush' he would do his best to get a fortnight in New Zealand and to return via 'Frisco'.[66] Here, immediately, is the energetic and enthusiastic traveller and teacher, restored to his more usual self by the prospect of escaping from the whole London world. It is the contrast between the tone of this letter and the earlier ones of 1904 which demonstrate so conclusively the depressing effect of that winter.

Later events show that the departure from London was fortunate.

awful winter, at least in London, and the fogs and damp never stop. The streets are indescribable.' In addition, or more probably because of all this, his health was beginning to suffer and for the only time in his correspondence with Rutherford he mentioned that he had 'not been very fit lately'.[57]

Despite these problems, there was one feature of this period which was to prove the beginning of a long and satisfactory relationship, although its consequences would have been less happy if Soddy and Rutherford had not managed to solve their differences over the 'Radioactivity' book. As soon as the first of the lectures had been published in *The Electrician*, 'their translation appeared in St. Petersburg in the form of a book as *Radioactive Phenomena*'. In the early decades of the century there was little or no primary research in radioactivity in Russia but an enormous appetite for information about developments in other countries was developing. To satisfy this, publishers in St Petersburg and in Moscow translated and produced editions of works considered 'suitable for broad circles of readers'. Amongst these, 'Soddy's book was especially interesting for Russian physicists and chemists' not only because its author had been involved in the initial discoveries, but also because of 'its visual clarity, definiteness, and beautiful popular style'. Beyond this, its particular attraction for Russian readers was 'the broad philosophical coverage of various problems of the relationships between matter and energy, problems which are of the utmost interest to Russian scientists.'[58]

In 1905, *Radioactivity. An Elementary Treatise from the Standpoint of the Disintegration Theory* was also translated, and after this, most of his work reached a wide audience in Russia. The most successful of these was *Interpretation of Radium* which was published in six different forms in Russia. The reason for this popularity was that[59]

Soddy's work can be singled out from the numerous books written about radioactivity by the nature of its presentation. The author always remains on a rigorously scientific plane ... but at the same time in his presentation he can employ not only uncommon simplicity, clarity and briskness but also the breadth of a philosophical approach and the touch of artist taste. Even when the author is carried away by his imagination, which is quite noticeable at times, one can sense the spontaneity and frankness of the original creative thought.

This association resulted in Soddy's election to member–correspondent status in the USSR Academy of Sciences in 1924 through which his particular skills in communicating complex scientific theories and experiments was recognised.

However, in 1903 this was far in the future. Immediately, the result of the difficulties in the other aspects of his life was that Soddy was increasingly disillusioned. By December 1903 he felt very despondent:[60]

I see my sphere of usefulness at University College is accomplished and shall sitting in the front row listening as diligently as any student. I felt as a classical don might if Homer had dropped in to hear him lecture on the *Iliad*.'[50]

Soddy also gave a number of lectures to lay audiences as it had 'always been my conception of part of the duty of the academic scientist to try to impart as much as possible of his knowledge to the layman.'[51] He wrote popular articles like that in *The Contemporary Review*; the lectures on 'Radioactivity' were published as a series in *The Electrician*, and it was proposed that they should also provide the basis of a book on the subject. This led to a dilemma for Soddy. He had agreed that Rutherford, as the senior partner, should publish the first complete results of their collaboration. Now, because he was breaking their agreement, he felt he needed Rutherford's consent before authorising the book. On the other hand, the publishers of *The Electrician* 'have decided to have a book on Radioactivity in their Electrician series. I have an excellent opportunity of writing it now as I give my lectures.' If he refused they would publish one anyway, by plagiarising Soddy's work if necessary.[52] Rutherford was furious. Should Soddy go ahead, his book 'cannot help being a fractured replica of mine' and would be a 'bad breach of scientific etiquette'.[53] It took all Soddy's skill to persuade Rutherford that the popular book would be published no matter what they did and so the only control they could exert would be if Soddy wrote it. He, then, could delay publication until after Rutherford's more detailed and learned account.[54] It is a reflection of the strength of their friendship that, a month after this furious exchange, they were both writing normally again about scientific gossip and their work. The book was published as *Radio-Activity* early in 1904 and, as predicted, was a popular success but never proved serious competition for Rutherford's later, authoritative account.[55]

In the meantime there were other complications connected with Ramsay's behaviour. Rutherford and Soddy had never hoped for individual monetary gain from their work although they both wanted more resources for their laboratories and at various times hoped for larger salaries. Ramsay seems to have held a very different set of beliefs and by the end of 1903, Soddy was wondering whether Ramsay was giving interviews to the press because he 'foresees the great commercial advantage in the future of being known as the expert on radium'.[56]

There were other problems as well. Various uncongenial elements in Soddy's life were seeming to conspire against him. He was under some great pressure of work, and finding the lecture series for *The Electrician* 'a frightful bind', a very unusual complaint from one who always responded so well to an audience. He was also beginning to feel the strain of the London winter. Although he disliked hot weather, his preference was for the cold clear snowy winters of the mountains while 'we have had an proved that one element could be changed into another, is that I was entirely absorbed in the complicated gas manipulation . . . as I worked, the room began to fill silently, as by some telepathic process the news of the success of the experiment began to spread through the laboratory. . . all standing space was occupied . . . the spectroscopes passed form hand to hand . . . I suppose I was the last in that room to see that single bright yellow line which Lockyer had discovered in the sun's chromosphere thirty-four years earlier . . .

This discovery was the 'chemical sensation of the summer of 1903', but also possibly a salutary experience for someone who, only three years earlier, had asserted the impossibility of particles smaller than the atom, and had poured scorn on the theories of the same Lockyer. It is a pity, if not surprising, that Soddy's feelings on this remain unrecorded.

Although it appeared that all was going well, there were problems through this period. The first was related to his work. The relationship with Ramsay began with some sense of strain. Soddy and others were concerned about Ramsay's work at the time and there were faint stirrings of rumours of plagiarism connected with Ramsay's writing. Soddy was anxious that the final Rutherford/Soddy results should be published so that Ramsay and others would not 'make asses of themselves'.[47] There is also some suggestion that Ramsay was reaching the end of his time as a successful chemist and that, while his name was still powerful, he was beginning to acquire a reputation for inaccuracies which would rub off on to his fellow workers.[48] Most importantly, there seems to have been some disagreement over a point of principle about publication of results. This was a constant problem for researchers. When groups and individuals in various countries were working on similar questions the credit for any breakthrough was given to the first in print. In an age of slow international communications, sometimes unreliable post and protracted publishing times the race for publication could be intense. There was also the difficulty associated with publishing partial results. Once in print, some credit at least would be awarded, but the article might point the way for others to find the correct solution first.[49] For Soddy these were very real difficulties in 1903.

He could do nothing except worry in the case of Ramsay, but was in a more urgent predicament with Rutherford. Although Rutherford and Soddy had violent arguments, they both accepted that scientific discoveries should be published freely for the benefit of anyone who could make use of them. They had also initially agreed that Rutherford, as the senior partner, should write the first, academic account of their work. However, Soddy had found that on his return from McGill almost everyone wanted to know of the discoveries there. He addressed specialist audiences; he gave twelve lectures on 'Radioactivity' at University College, London and a shorter series at the Cavendish at the invitation of J. J. Thomson where 'it was somewhat embarrassing when I saw that the great man himself was

While the whole McGill experience had been enormously enjoyable, both socially and professionally, it was short lived. By early 1903, the chemical part of the experiments was almost complete and the direction that Rutherford intended to pursue was one for physicists. Soddy had decided on a line of research more suited to a chemist—the exploration of the rare gases and their possible radioactive powers. Sir William Ramsay had perfected the appropriate experimental techniques and although no definite arrangements for the future had been made, by October 1902 Soddy had given up his post as demonstrator and arranged to return to England, intending to work with Ramsay. In February 1903, Soddy boarded the *Merion* in Boston to sail for Liverpool.

After a brief visit to his family in Eastbourne, Soddy had an interview with Ramsay, whom he already knew. Ramsay had been the external examiner for Soddy's Oxford Final Examinations. Soddy made arrangements with Ramsay to begin work in his laboratory in London in early March. This was less than a month after leaving Canada.[43]

As usual, there were loose ends to be finished in the laboratory before work on the new research project—the gases given off in radioactive changes—could begin, so that it did not start until after Easter. Almost immediately, Soddy had a piece of good fortune almost as great as his decision to go to Canada. One of the greatest problems in any of the experiments into radioactivity was a purely practical one. There was a great shortage of the raw material, radium. It could only be obtained from the Curies in Paris who produced it through a long and tedious process of separation from pitchblende and so supplies were very limited. One result of this was the importance of Ramsay's particular skill, experiments with very limited quantities of gas, and it had also dictated the nature of experiments which could take place. The story is best told in Soddy's words:[44]

> Then, by the most extraordinary chance, the whole future prospect was changed. This seemed to me at the time like the direct intervention of Providence. I was walking along Upper Mortimer Street, off Upper Regent Street, in London one day when, casually looking in Isenthal's [a chemical supplier] window, I saw something I could not credit to be true: *Pure radium compounds on sale here.*

Suddenly, the radium shortage had come to an end. The commercial production of radium had at last been undertaken by 'Professor Giesel in Germany' from radium residues left after the extraction of uranium from pitchblende.[45] As a result, sufficient quantities of the radium compound were available and the experiments could begin.

They were successful almost at once. They had started work in March. As Soddy's account shows, in early July they were able to demonstrate transmutation:[46]

> My enduring recollection of the first successful experiment, in which it was

these points, he was never reconciled to the fact that many of the accounts ignore his contribution and speak of Rutherford as a chemist. Most infuriating of all was the Nobel Prize which was awarded for these discoveries in 1908. It was given to Rutherford and given for chemistry. Rutherford himself was amused by this and often referred to his rapid transformation, but Soddy was always bitter and his own prize, awarded in 1921, never fully compensated for what he regarded as this early slight.

The reasons for this lack of recognition are complicated, but, even at this early stage, one difference between Rutherford and Soddy is illustrated which was to have a fundamental effect on their futures. Rutherford was a pragmatic experimentalist, while Soddy, although a brilliant practical chemist, was also a philosopher. While Rutherford was concerned about the purely scientific implications of their work, Soddy was anxious to put the discoveries into a historical and social context, returning again to his recurrent concerns about alchemy:[40]

Nature can be a sardonic jester at times, when you come to think of the hundreds of thousands of alchemists in the past few thousand years toiling and broiling over their furnaces, spending laborious days and sleepless nights trying to transmute one element into another, a base into a noble metal, and dying unrewarded in the quest, whilst we at McGill, by my first experiment, were privileged to see, in thorium, the process of transmutation going on spontaneously, irresistibly, incessantly, unalterably! There's nothing you can do about it. Man cannot influence in this respect the atomic forces of Nature.

At the moment of the discovery, Soddy might have believed that Nature was all-powerful, but he was satisfied with this idea for only a very short time. Very soon he began to wonder whether the process could be harnessed and, even more radically, to consider the further possibilities stemming from calculations of the potential energy stored in the atoms. In a popular account of 'Some Recent Advances in Radioactivity' in 1903 he explained, 'atomic energy must be beyond all proportion greater than molecular energy, which is the source of the motive-power of our furnaces and boilers. There is no saying what strange results the recognition of this may lead to in the future.'[41] At this time, his prognosis of these results had been pessimistic, and must have been one of the earliest predictions of potential disaster stemming from atomic power. 'The knowledge [must] make us regard the planet on which we live as a storehouse stuffed with explosives, inconceivably more powerful than any we know of, and possibly only awaiting a suitable detonator to cause the earth to revert to chaos.'[42] He also suggested other uses for this amazing substance. In the *Times Literary Supplement* and the *British Medical Journal* in 1903 he suggested that radioactive substances might prove useful in curing disease, especially tuberculosis, and also in various industrial processes.

kind of exaltation . . . I remember quite well standing there transfixed as though stunned by the colossal import of the thing and blurting out—or so it seemed at the time: 'Rutherford, this is transmutation: the thorium is disintegrating and transmuting itself into argon gas'. Rutherford's reply was typically aware of more practical implications, 'For Mike's sake, Soddy, don't call it *transmutation*. They'll have our heads off as alchemists. You know what they are.'

The nine joint papers between 1902 and 1903 grew out of a period, from October 1901 to April 1903, of quite extraordinarily productive research springing from this one central and momentous discovery which was, at its most simple, the transmutation of an element.[37] They investigated the gaseous emanation from thorium and discovered that it was the effect of the disintegration of the thorium during which the intermediate but chemically separable substance, thorium X, was produced. This substance was very short lived so that its production was balanced by its decay. Thus the uncontrolled disintegration of one substance resulted in the formation of another. From this initial research, they produced a complete theory of radioactive disintegration which has remained more or less unchallenged, needing only a minor extension in 1909.

With this discovery they immediately came to the forefront of the scientists of their generation. At this point in their careers their futures seemed assured and fame, if not fortune, seemed certain. In the event, this was not the case. The collaboration between Rutherford and Soddy has been summarised by Feather as unique in three ways: for the extreme youth of Rutherford and Soddy, the entire novelty of the field and 'the fact that over a period of thirty years or more a fair share of credit for their joint achievement was not universally accorded to one of them'.[38] This estimate is perhaps rather overstated. Rutherford and Soddy were young, but Rutherford was thirty, which is not really 'extreme youth', and others were very active in the field of radioactivity if not in the specific aspect chosen by Rutherford, that of the emanations from thorium. For us, the third part of the summary is more important. Both Soddy and Rutherford always insisted that their discoveries were entirely joint products. They required the combined knowledge and skills of a physicist and a chemist, perhaps this particular pair. As Wilson has commented, '. . . It was not only his chemical expertise which complemented Rutherford's physics . . . Soddy was the clever member of the pair where Rutherford provided the power; Soddy was much the quicker of mind and much the better writer, lacking Rutherford's penetration but better at seizing the wider philosophical implications of their joint work.'[39]

However, Rutherford was awarded most of the credit for the results of their work. In some ways this was understandable since he was the senior partner, his was the initial idea, and the discoveries were made in his laboratory. Nevertheless, while Soddy would have agreed with many of of matter is the province of chemistry, and little indeed can be known of this constitution until transmutation is accomplished . . . This is . . . the real goal of chemistry . . .

This paper has two areas of importance for any understanding of Soddy's future work. Firstly, it shows the first sign that he was accepting the possibility that atoms were not unchangeable. This was the whole basis of his collaboration with Rutherford—the idea of penetrating into the construction of the atom. By implication, if the atom consisted of some kind of smaller bodies, then there was at least the theoretical possibility of altering the constituent parts, of altering the atomic structure. Once the atomic structure was changed, then the atom could be changed and the Daltonian and Newtonian theory of the indivisible atom would finally be challenged. Even to embark on this work, Soddy would have to alter his views and move from the position he had so publicly taken in debate. However, as a supreme rationalist, once convinced of errors in his thinking, Soddy could change his opinions with often violent speed.

There was another, philosophical, element in the results of Soddy's researches for this paper which was to have reverberations throughout much of his life. Here he suggested the possibility of a history of human powers over nature which had been disguised as magic, thus unwittingly prophesying the reactions which were to greet the discovery of radium. He also refers to the use of the symbol of Ouroborous, the tail eating serpent, in manuscripts from the 'III and IV centuries' to express the cyclical nature of matter. This notion was to be almost his only support when he came to realise in the mid-1900s the full implications of his work with Rutherford. A cyclical world might have succumbed to disaster in past aeons, he was to suggest, but had managed to overcome the disaster and return to a degree of comfort and civilisation.[33] There is some disagreement about the actual date of this paper.[34] However, internal evidence seems to suggest that it was written either before or at the very beginning of the Rutherford–Soddy collaboration, since it contains the basic ideas needed for consideration of the problems they were to investigate.

This problem centred on the nature of what they termed an 'emanation' from the element thorium. Their achievement was to show experimentally that this emanation was the product of a constant change of the element into the chemically separable thorium-X and the emanation. Radiation from the thorium was shown to be particulate and a direct accompaniment to the process of disintegration.[35] When the first undeniable evidence of change was measured in the laboratory, they reacted to the discovery in their various ways. Soddy was[36]

> overwhelmed with something greater than joy—I cannot very well express it—a

were many advantages to living in Canada, there were also some serious disadvantages stemming from the distance from McGill to the centre of scientific discoveries in Europe. As Rutherford described the situation during his time at McGill:[26]

After the years in the Cavendish I feel myself rather out of things scientific and greatly miss the opportunities of men interested in physics. Outside the small circle of the laboratory it is seldom I meet anyone to hear what is being done elsewhere. I think that this feeling of isolation is the great drawback to colonial appointments, for unless one is prepared to stagnate, one feels badly the want of scientific intercourse.

Not only were there problems about meeting other scientists, but communications of all kinds were slow so that news of discoveries made by others and the publication of any articles was delayed. This could prove a serious disadvantage for a researcher who constantly felt the pressure from others in the field. As Rutherford wrote to Mary, his future wife, 'I have to keep going as there are always people on my track. I have to publish my present work as rapidly as possible in order to keep in the race.'[27] This was to prove even more of a problem when experiments were successful. Thus, when Soddy returned to London he found that as their final papers were still to be printed he was unable to speak to his fellow workers for fear of revealing these results.[28]

Their sense of isolation in Canada caused both Soddy and Rutherford to apply for other jobs. At the end of 1900, Soddy learnt that the Chair of Chemistry at Aberystwyth was to become vacant and, having confirmed that there was little likelihood of promotion at McGill in the immediate future, he sent off his testimonials.[29] Rutherford, on the advice of his mentor, Thomson, tried for the Chair of Physics at Edinburgh University that was vacated in March 1901. He did not really expect to be appointed since he had been warned that there was serious competition for the post, but wanted to make a gesture to inform the rest of the profession that he would be prepared to move if a suitable offer was made.[30] Both attempts were unsuccessful and so Rutherford and Soddy settled down to life at McGill.[31]

During this period between the debate and the beginning of actual research, and possibly as a response to Rutherford's contribution to the debate, Soddy went back to a study of the written material on the whole subject of radioactivity and transmutation. The result of his reconsideration of the possibility of the alchemical predictions being correct and of a connection between alchemy and chemistry are summed up in an undated paper which he gave at McGill. Here, he voiced his doubts:[32]

The existence of the atomic stage of matter is unquestioned, but this in no way relieves the chemist from the necessity of penetrating deeper. The constitution of the distinguished astronomer, Lockyer, saying, 'Astronomers often appear to lose themselves in the contemplation of the celestial, and such mundane considerations as logic sometimes escape them.' While he did not attack Rutherford personally, he managed to suggest that he was some kind of foolish visionary who could not recognise reality, but was dealing 'in a new world demanding a physics and a chemistry of its own'. He concluded with a statement of his own position: 'I feel sure that chemists will retain a belief and a reverence for atoms as concrete and permanent identities . . . if transmutable, certainly not yet transmuted.'[20]

Rutherford gave his prepared answer but, as Soddy reported the scene later, 'he appeared to be completely taken aback at the vigour of my onslaught and I saw that I had hurt him bitterly before a host of his own students.'[21] Rutherford was not the only person to feel that Soddy's attack was rather violent and more suitable to the Oxford Union where he had learnt his style than to the more sedate McGill. The audience generally felt that he had overstepped some tacit mark, especially since he was the junior and the visitor. He almost made an enemy where he needed a friend. Rutherford was unused to being forced on to the defensive, he 'was very impatient of anyone who discoursed at greater length than he liked, and when he wanted to take the floor himself; on such occasions he was frequently quite ruthless and thoroughly trounced the offender'.[22]

The debate had two results. Firstly, although he was demonstrably in the wrong, Soddy was judged to have carried the day. Secondly, and much more importantly, despite his manner, the content of his contribution impressed Rutherford. As a result, and despite all their differences, he then asked the young chemist to join in 'some investigations' he was about to start. Flattered at the suggestion and already somewhat bored with the ordinary tasks of teaching, Soddy was delighted to accept: 'After months of routine chemical work the thought of getting on with some original investigation gave me such a mental stimulus that I found some difficulty, on going back to my own laboratory, in settling down . . .'[23] He had found the 'virgin soil' to which he had referred in the Evans Prize Essay he had written in 1895, 'on the borderland . . . between chemistry and physics . . .'.[24]

Rutherford and Soddy began planning for this research soon after the debate in March 1901, but the actual work was not to start until the October. The everyday tasks needed to be completed and then the summer vacation intervened. Rutherford spent his summer with his wife and their new baby daughter in comfortable domesticity. In contrast and typically, Soddy returned to Britain and went to the meeting of the British Association in Glasgow.[25]

The need to travel back to England to keep in touch with new advances in the subject demonstrates the extent to which, while there this Rutherford stood not only as a physicist but as someone who had been supervised by Thomson himself at the Cavendish Laboratory. This introduced not only a disciplinary difference but one of personal allegiance and finally of university loyalty.

In addition to disciplinary and other differences, the personalities and backgrounds of Soddy and Rutherford were opposed. Although Rutherford presented a picture of an amiable, bucolic and powerful figure in his successful later years, 'at this time of his life [he] gave quite a different impression . . . he could be subject to black despairs and monumental rages, to enormous enthusiasms and to brilliant public exhibitions.'[17] Soddy described his first view of Rutherford like this: 'I heard a resounding voice echoing down the corridor and a powerfully built young man strode down the passage and into the laboratory. He had all the characteristics of a Colonial and I knew it must be Rutherford . . . his gauche manner made me wonder how such an energetic person could possibly avoid smashing the delicate instruments which he would have to handle. It looked to me like a case of a "bull in a china shop".'[18] In contrast to this, Soddy, although tall, was never large. He had long, delicate hands well suited to the tasks of a chemist. He was precise in his movements and careful in his dress. He tended to be quiet, inward and withdrawn, and his black moods were manifested as sarcasm and a particularly biting kind of wit, which had been sharpened by his years at Oxford. Perhaps all the differences were contained in Soddy's description; Rutherford was a Colonial while he was English—the new versus the old, or, in the terms of the debate, the modern against tradition. All these incompatibilities were demonstrated in the debate, in which each felt he was defending the honour of his discipline against an antagonistic opponent.[19]

Anticipation of this confrontation of different scientific beliefs attracted a large audience. Rutherford had announced that 'we hope to demolish the chemists', a challenge which Soddy could not resist. He began his paper with the announcement that he had been asked to present the views of the 'conservative' chemists,

in the malicious anticipation, I have some reason to suspect, that I shall thereby deliver myself and my case over, bound into the hands of the adversary or shall at least afford an easy target for the modern artillery at the disposal of the ionists.

Even allowing for the depth of feeling involved, Soddy's presentation of his case was extreme. Whether he was carried away by the occasion or was reacting to a perceived threat to the theories that were the basis of his chemical knowledge is unclear, but he attacked and simultaneously refused to believe the threat was real. He berated the physicists; he accused Thomson of muddled thinking; he condemned the findings

Some of this was more engaging than just setting up and explaining experiments. Particularly, a lecture series on the history of chemistry awoke his enduring interest in the subject and he was to return to it several times in the future. It was this area that was to provoke his unease about the possibly cyclical nature of scientific progress, coming eventually to suggest the possibility of an earlier generation of scientists who had destroyed their world through uncontrolled scientific experiment.[13] However, this was in the future. In these first lectures given in the autumn of 1900 Soddy described the history of his discipline, beginning in ancient Egypt and finishing with a description of the Daltonian theory of matter, arguing for the immutable and indivisible atom. He poured scorn on the alchemists' ambition of changing lead into gold, insisting that any notion of the possible transformation of the elements from one to another was impossible. However, he was clearly fascinated by the whole question and had read widely about it. He used the alchemical term, transmutation, for the process and devoted two lectures to the practice of alchemy, illustrating them with slides of alchemists and their work.[14] However, he revealed his distance from the practices by asserting that 'the alchemistic period had nothing to do with the normal development of chemistry. It is rather the result of a mental aberration.'[15] While interesting, these drafts of his early lectures illustrate Soddy's ability to choose the 'wrong' side in an argument and to refuse to listen to alternative views. His obstinacy was demonstrated in a public debate organised by the McGill Physics Society on the subject 'The Existence of Bodies Smaller than the Atom', held on 28 March 1901. The speakers were to be Soddy, defending the point advanced in his lectures by speaking against the motion, while Ernest Rutherford was to defend the proposition.

At the time, the subject was of great topical interest. The basis of all nineteenth-century science (especially for chemists) was Dalton's suggestion in 1803 that the basic constituent of matter was the atom; that each element consisted of enormous numbers of identical atoms, and that the particular atoms of the various elements were different from one another.[16] This theory remained unchallenged until the 1890s when J. J. Thomson had discovered that cathode rays were composed of negatively charged particles and not, as had been thought, of waves. He announced the discovery of these 'electrons' at the Royal Institution in 1897. The electrons were calculated to be only one thousandth the mass of the lightest atom and were identical, no matter what substance they came from. This discovery had shaken the whole intellectual world since it changed the whole basis of theories of matter. But there were sceptics who were not convinced by the evidence. Since Thomson was a physicist, chemists numbered large amongst the sceptics, and amongst them was Soddy who continued to support the Daltonian view. Against his holiday with the Carpenters, the scenery here was impressive and, on occasion, dangerous. On one of his trips Soddy had the 'closest shave I have ever had climbing'. This occurred while attempting to climb Mt. Abbot. His chance companion having failed to climb a 500 foot cliff, Soddy went on alone, managed to reach the top and then, on the descent, misjudged the distance to a ridge and 'I had to let go . . . fell on a sloping bit . . . and down I went . . . over and over, clawing at the earth, to come to a stop on the very edge of the 500 foot drop.' This experience was not enough to quell his enthusiasm and throughout his life one of his favourite leisure activities was climbing in mountains, of Wales or the Alps. The essential emptiness of mountains, especially in snow or rain, provided some relief from everyday contacts for this solitary character.

He was also intensely curious about Canadian society and so took the opportunity to travel. For example, he went to the fair in Portland where 'it was a wild night' and the police had to be called in to restore peace and order. In Ottawa, while waiting in a hotel room for his train, 'a blast of flame shot up outside the bedroom window'. The building next to the hotel had caught fire. As he noted, this was not unusual in Canada at the turn of the century. Timber was stored in the centre of Ottawa so there were several devastating fires, while in Montreal the combination of the steam or hot water heating in the houses and the dryness of the air from an air temperature below zero Fahrenheit caused the woodwork to become intensely flammable and so 'I once saw a quarter of a square mile of blazing roofs . . . and firemen getting literally frozen to their ladders'.[11]

While Canada provided both friendship and memorable experiences, Soddy retained the awkwardness in unfamiliar situations or when meeting new people which had marked his time as an undergraduate. This was demonstrated by what was to prove perhaps the most consequential meeting of his life. The Junior Professor of Physics at McGill, Ernest Rutherford, had been in his native New Zealand marrying his childhood sweetheart, Mary, when Soddy arrived at McGill. They did not return until the autumn of 1900 and so missed Soddy's first six months. On his return, Rutherford needed to gather the reins of the department again after his trip and then he started a new series of experiments. Meanwhile, Mary Rutherford was settling down to life in a new country, taking on the mantle of faculty wife, 'as hostess at regular meetings of the research students in their home, playing her full part in supporting her husband and his work'.[12]

During this period, Soddy had been busy preparing lectures and experiments for his students. While he enjoyed the work, he also found it exhausting, discovering that the post of demonstrator was no sinecure. He was expected to arrange and supervise experiments, to produce a new course on gas analysis and to lecture to the students.

his home, assuring him that 'Toronto could never spot a winner', and partly by Soddy's desire to see more of Canada. Deciding to maximise his opportunities while in the country, on his way home Soddy stopped at Montreal to visit the MacDonald Laboratories at McGill University. These had been provided by the munificence of the anti-smoking tobacco baron, Sir William MacDonald, and were known to be amongst the best of their kind at the time.[6]

On arrival, Soddy was met by Professor Bernard James Harrington, the Professor of Chemistry and Director of the MacDonald Chemistry Building, who showed him around and was persuaded by Soddy of his suitability for the post of demonstrator, which was then vacant, in the department. This offer was made not only on the basis of Soddy's fitness for the post. The junior professor in the department, William Wallace, was in England getting married and, at the same time, finding a suitable applicant for the post. It amused Harrington to pre-empt his junior's efforts in this way, and so he appointed Soddy. The salary of £100 a year was far below the £400 that had been promised as a probable salary by Hughes way back in Eastbourne, but it was the first salary that Soddy had been offered, and was to be preferred to an ignominious return to England.[7]

Soddy accepted the offer and this was to prove one of the most propitious decisions that he ever made. One contributory factor to his success in Canada was 'the cordiality with which I was received and treated by a very remarkable group of men then at the university.'[8] Soddy was always to react to the warmth of the esteem in which he was held by friends and colleagues. His success depended on a background of support and friendship and when, as at McGill, this was provided, his professional achievements were outstanding. Here his associates gave the much publicised North American welcome, which extended beyond their working environment to entertainment during the vacations. For example, [9]

Bovey was Dean of the Science Faculty and he and his family entertained me at their summer house in Little Metis on the Lower St. Lawrence one summer vacation; I spent a delightful and profitable time there . . . amongst kindly, human and picturesque surroundings, an atmosphere very much to my liking.

There was a whole range of other similar events: a trip to the backwoods, sleigh rides in the winter, horse riding in the summer, bathing in the river and exploring the lakes. He was able to take photographs with the camera that he carried on all his journeys, and he kept three of these photographs on his study wall for the rest of his life.[10]

While he was in Canada his love of hills and climbing grew. Compared with the English Lakes, where this affection had been awakened during

CHAPTER 2

Success

The years from 1900 to 1904 were to be the most successful of Soddy's career. His later achievements stemmed from this early period and the colleagues and friends he made at this period were to influence the whole of his life. However, there was no intimation of this during his first days in Canada.

After his somewhat precipitate departure from Oxford and an uneventful sea crossing to New York, Soddy boarded the train for Toronto. He took with him the daily paper in which there was a report of speeches given at the farewell dinner for Dr Pike, the retiring Professor of Chemistry in Toronto.[1] In this report, Soddy found words which dashed whatever hopes he had entertained for his application for the chair. In his speech, the Principal of the university had said, 'Dr. Pike came to us from Oxford, but thank God his chemistry was made in Germany'.[2]

This was a reference to what were seen as two diametrically opposed styles of chemistry during the late nineteenth century. The Germans had developed the natural sciences as a necessary and natural adjunct to the technological advances being made in her manufacturing industries while in England there was still a tendency for science to be seen as the province of the gentleman amateur. As we have suggested, this position was changing, but the change was not yet widely recognised. Certainly Toronto seemed unaware of it. For Soddy this pro-German feeling was a blow. He had, anyway, 'stood little chance of success: he was, after all, not twenty-three years old, with no teaching or administrative experience and only the merest smattering of research training',[3] but youthful optimism together with a natural impetuosity had propelled him so far. This tendency to act before thinking was to remain with him. As one of his obituary writers said, 'no-one can doubt that he was a strange man whose own impetuous actions created most of the difficulties that beset his life.'[4]

However, on this occasion, while perhaps it was ill-judged, his journey to Canada was to prove fortunate. His first reaction to the news of the anti-Oxford feeling in Toronto had been to begin 'to make preparations to get back to England as soon as possible' but 'though Toronto did not gather me to its arms I was destined not to leave Canada for nearly three years.'[5] This delay was caused partly by the kindness of Irving Cameron, the Professor of Surgery in Toronto, who took the young man to stay in

38. Professor Frederick Soddy Papers, Bodleian Library, Oxford, MS Eng Misc b180.
39. Soddy, F. (1898). The Life and Work of Victor Meyer. *Transactions of the O.U.J.S.C.*, May 1898.
40. Cameron, N. (1979). 1900: The Cavendish Physicists and the Spirit of the Age. In *Rutherford and Physics at the Turn of the Century*, (ed. M. Bunge and W. R. Shea), pp. 124–39. Dawson, New York.
41. Memoirs, p. 21.
42. Pioneer, pp. 45ff.
43. *Memoirs*, p. 21.
44. *Memoirs*, p. 27.
45. Hartley, H. (1965). The Contribution of the College Labs. *Chemistry in Britain*, 1, 22.
46. *Pioneer*, p. 48 and p. 52, where the testimonial is quoted.
47. *Pioneer*, p. 50.
48. Soddy, F. (1933). Reminiscences of McGill, 1900–1902. *Old McGill Annual*, **1936**, 17.

14. Personal communication from Mr Robin Harrison, Secretary of the Old Eastbournian Association.
15. Smith, D. C. (1986). *H.G. Wells: Desperately Mortal*, pp. 10–12. Yale University Press, Newhaven.
16. Perkins, H. (1989). *The Rise of Professional Society: England since 1880*, p. 84. Routledge.
17. Personal communication from Mr Robin Harrison, Secretary of the Old Eastbournian Association.
18. Professor Frederick Soddy Papers, Bodleian Library, Oxford, MS Eng Misc b170.
19. The only people connected with the school who had any recollection of Soddy were the retired librarian, Mr Brian Harrap, and Mr Robin Harrison, and the only memorial is a very dusty plaque in a corner of a laboratory. See *Nature*, 1 November 1958, **182**, for an account of this memorial.
20. Soddy, F. Some Personal Memories of the Earlier Years of Harold Carpenter's Career. Typescript in Professor Frederick Soddy Papers, Bodleian Library, Oxford, MS Eng Misc b180.
21. Quoted in Roderick, G. W. and Stephens, M. D. (1972). *Scientific and Technical Education in Nineteenth-Century England*, p. 30. David and Charles, Newton Abbot.
22. Professor Frederick Soddy Papers, Bodleian Library, Oxford, MS Eng Misc b180.
23. Hughes, R. E. and Soddy, F. (1894). The Action of Dried Ammonia on Dried Carbon Dioxide Gas. *Chemical News*, **22**, 3.
24. Quoted in Roderick, G. W. and Stephens, M. D. (1972). *Scientific and Technical Education in Nineteenth-Century England*, p.40. David and Charles, Newton Abbot.
25. Smith, D. C. (1986). *H. G. Wells: Desperately Mortal*, Chapter 1, *passim*. Yale University Press, Newhaven.
26. Wiener, M. (1981). *English Culture and the Decline of the Industrial Spirit, 1850–1980*, p. 132. Cambridge University Press.
27. *Pioneer*, p. 36.
28. *Memoirs*, p. 17.
29. *Memoirs*, p. 17.
30. Frederick Soddy, 1895, quoted in *Pioneer*, pp. 36–7.
31. Krivomazov, A. N. (1986). The Reception of Soddy's Work in the USSR. In *Frederick Soddy (1877–1956): Early Pioneer in Radiochemistry*, (ed. G. B. Kauffman), p. 132. Reidel, Dordrecht.
32. Professor Frederick Soddy Papers, Bodlein Library, Oxford, MS Eng Misc b170.
33. *Pioneer*, p. 39.
34. Professor Frederick Soddy Papers, Bodleian Library, Oxford, MS Eng Misc b170.
35. Professor Frederick Soddy Papers, Bodleian Library, Oxford, MS Eng Misc b180.
36. Professor Frederick Soddy Papers, Bodleian Library, Oxford, MS Eng Misc b180.
37. Professor Frederick Soddy Papers, Bodleian Library, Oxford, MS Eng Misc b179.

all kinds of extravagance which his family could not afford. The hope was that a long sea voyage to Italy and back would cure him. The effect on the student is unknown, but the episode left Soddy feeling 'like a Boy Scout who has done his good deed, not for a day, but for a life time'.[47] Perhaps it also afforded him time to consider his own position for, when he returned, he changed the course of his life. He heard that a Dr Pike was to vacate the Chair of Chemistry at Toronto University. In a typically impetuous action, almost without a second thought, as far as can be ascertained without consulting his teachers or family, he applied for the post. His only explanation for this action was that the retiring professor 'had been the predecessor of my chemistry tutor at Merton College, Oxford, Dr. John Watts'. As he continued, 'not a very good qualification perhaps, but an excuse for a visit to Canada, which again needs no excuse'.[48] Then, incredible though it seems, without waiting for an answer from Toronto, he packed his bags and his packet of testimonials and set off for Canada.

Notes and references

1. Interview between the author and Marjorie Soddy, Frederick's niece, June 1991.
2. Professor Frederick Soddy Papers, Bodleian Library, Oxford, MS Eng Misc b170. (Not all this material is foliated or numbered.)
3. Quoted in Cannadine, D. (1980). *Lords and Ladies: the Aristocracy and the Towns 1774–1967*, p. 230. Leicester University Press.
4. East Sussex Record Office, Lewes. NMB 15/1/1 Trustees' Minutes, Eastbourne Wesleyan Methodist Chapel.
5. Trenn, T. J. (1975). Frederick Soddy. *Dictionary of Scientific Biography*, (ed. C. G. Gillispie), pp. 504–9. Charles Scribner's Sons, New York.
6. Russell, A. S. (1956). F. Soddy, Interpreter of Atomic Structure. *Science*, **124**, 1069–70.
7. Soddy, F. (1920). *Science and Life*, p. 205. John Murray. Hereafter, Soddy's works are referred to by a short title, followed by a page number.
8. East Sussex Record Office, MS Census of Population, 1881 Census, Enumerator's Books, RG 1040.
9. Howorth, M. (1958). *Pioneer Research on the Atom: Rutherford and Soddy in a Glorious Chapter of Science: The Life Story of Frederick Soddy, M.A., LL.D., F.R.S., Nobel Laureate*, p. 29. New World Publications, London. (Hereafter *Pioneer*.)
10. Professor Frederick Soddy Papers, Bodleian Library, Oxford, MS Eng Misc b170.
11. Howorth, M. (1953). *Atomic Transmutation: The Greatest Discovery Ever Made from Memoirs of Frederick Soddy M.A., LL.D., F.R.S., Nobel Laureate 1921*, p. 17. New World Publications. (Hereafter *Memoirs*.)
12. Professor Frederick Soddy Papers, Bodleian Library, Oxford, MS Eng Misc b170.
13. Paneth, F. A. (1957). A Tribute to Frederick Soddy. *Nature*, **180**, 1085–6.

in improvements to the equipment in the University Laboratories in the Museum so that students after Soddy's period were able to gain considerable manual dexterity in the manipulation of apparatus, a necessary part in any later research career. In the early twentieth century little laboratory apparatus was commercially available so specialist tubes and other appliances were made *in situ*. Skills in glass blowing and soldering were therefore an essential part of any scientific education.

Following his graduation, there was a somewhat disappointing hiatus in Soddy's career. A young man with a First at Oxford could well have expected some kind of fellowship, especially in an area of rapid growth like the natural sciences. Soddy has left no record of his expectations, but his lack of applications, to other universities or elsewhere, suggests that he anticipated some such offer. Instead, like many other graduates at the university, although he had no recognised post or standing, he spent the period from his finals in the summer of 1898 until the winter of 1899 in Oxford. Here he spent his time '. . . perhaps unwisely, attempting various lines on my own, having, with a Balliol graduate of my own year . . . fitted up a cellar on the then No. VI staircase as a Research Lab.'[44] It is possible that this is the same work as that referred to in a letter in 1899 when he wrote that he was working 'in the Balliol and Trinity laboratories . . . making one half of Marsh's formula of camphoric acid and Harold Carpenter is making the other half in Manchester'.[45] His aims during this period are unclear. Muriel Howorth, who spoke to him in his old age about this period, reports his feelings of uncertainty, '. . . what kind of job, and how to go about getting it? After all, perhaps he would not make that decision yet awhile. He would hang on for another year or so doing research . . . writing of papers and essays . . . prepar[ing] lectures and do[ing] some teaching.' Unfortunately, there are no reports of his teaching and lecturing so the worth of what he did here cannot be assessed. Howorth has suggested that Soddy remained at Oxford to complete some research which he had undertaken with J. T. Nance of Balliol, but gives no indication of the subject or results of that research, and this is supported by his testimonial from Sir John Conroy, FRS, Fellow and Bedford Lecturer of Balliol College, Oxford.[46] It is a hard fact of academic life that the only recognised method for measuring the value of research is by publication of results, and he published nothing. By these criteria he seems to have been wasting his time. Perhaps his later silence about this episode suggests that he recognised the lack of results.

This interpretation is strengthened by his uncharacteristic behaviour that summer which he spent on a voyage to and from Naples. This was the result of a series of events more suited to a melodrama than the life of a young scientist. Soddy was asked to accompany an undergraduate who had 'become deranged', with the result that he was indulging in

This had resulted in the late nineteenth-century attempts to improve the teaching of science generally in England. More particularly the Clarendon Laboratory in Oxford had been founded with money originally intended to provide a riding stable, and the Cavendish Laboratory in Cambridge with funds provided by the Duke of Devonshire. Until the late nineteenth century, science had been regarded as an almost entirely theoretical subject whose practical applications would be learnt after graduation, if at all.[40] Practical, experimental science was seen to be the province of the gentleman amateur, making his own discoveries and providing spectacular effects to amuse his visitors. As an increasing emphasis on practical, experimental skills grew, laboratory-based research institutions developed within the universities. However, the source of funding— private individuals and not the state—points to a continuing tension between the aristocratic ideal of the amateur scientist and the newer notion of the scientist as a professional. This tension was especially marked at the Clarendon, where the amateur ideal lingered into the twentieth century with disastrous effects on teaching and the building of research departments.

Despite these improvements, Soddy felt that the teaching of science during his undergraduate years in the university laboratories was 'almost incredibly bad, or rather, almost non-existent'[41] so he also attended the classes in the much superior college chemical laboratories in Balliol and Trinity. The division in the teaching of science between the university and the colleges was to persist at Oxford until the mid-twentieth century, and it was never ideal. The colleges had insufficient resources for adequate laboratory provision, but had the first claim on the loyalty of their students and dons. The university could provide the laboratories and lecture theatres but without the support of the faculty, the teaching within them could never be of the highest quality. This was a problem which confronted Soddy again when he returned to his Alma Mater as a Professor of Chemistry in 1919 and he found it as difficult to deal with then as he had when an undergraduate. He came to feel that one of his greatest failures within Oxford was his attempt to reorganise the teaching of chemistry.

Notwithstanding his complaints about almost all aspects of Oxford life, Soddy graduated with First Class honours in 1898 and a characteristically 'Oxford' introduction. At his Final Examination, he met William Ramsay for the first time, and they formed mutually satisfactory impressions of each other. They were to meet again when Soddy joined Ramsay's laboratory in London in 1903.[42]

Away from the personal, but directly echoing Soddy's experience, Ramsay's report as external examiner to the chemistry degree was highly critical of the experimental facilities available to students.[43] This resulted was far more of a holiday with no academic content, but during this holiday Soddy first began to know and love mountains and especially the Lake District, which was to become a lasting passion. Harold Carpenter's uncle, Dr Estlin Carpenter, was closely connected with, and later Principal of, Manchester College, Oxford. Carpenter and his 'sylph-like consort' habitually rented a house in Borrowdale for the summer vacation. They would take parties of students there, 'supplementing their university studies with practical inculcation of the love of Nature in the tradition of the great Lakes' poets.'[36] Harold Carpenter and Soddy joined one of these parties and the immediate, if not totally happy, influence of this experience can be seen in Soddy's verse composition for the Sir Roger Newdigate Prize for Poetry in 1898. This consists of 266 lines on 'The Pilgrim Fathers' in the high Romantic style of 'Yet murmur not, O Muse of Love'.[37] Fortunately, although the effects of the scenery resulted in a love of mountains and hills that he retained throughout his life, the effect on his literary style was short-lived.

The Carpenters spent a large part of their time in the lakes walking and climbing with the group of students and one memorable day took Soddy to see the sun rise on Sca Fell Pike. They breakfasted in Wastdale, lunched in Eskdale and then began their return via Esk Hause. Demonstrating the lack of a sense of direction that was more than once to lead him into trouble, Soddy 'innocently remarked . . . that we might as well come home by Glaramara'. Dr Carpenter took this as a challenge, and they managed the whole trip, but as Soddy ruefully remarked, 'it was I, the youngest of the three, who petered out as we got to the bottom, and they had to wait for me.'[38]

While Soddy later felt that Oxford was perhaps unsuccessful in terms of his social life, he does not seem to have been lonely. During his time as a student, he was a member of the Oxford Union and of the Junior Scientific Club, in whose *Transactions* his first individual paper was published in 1898.[39] Through these societies he gained experience in public speaking and especially a particular kind of scathing wit which could be seen as amusing to an undergraduate audience but was sometimes hurtful to those directly addressed. This was to be illustrated by his meeting with Ernest Rutherford, his future collaborator, when they both took part in a public debate.

Soddy was even more disappointed with his education in chemistry than with the social life at Oxford. He felt that there was a lack of opportunity in the courses which the university offered to the young science students. In this he was at least partly justified. Teaching of science was still not very good even by standards less exacting than Soddy's. Some progress had been made since the 1867 International Exhibition had revealed the threat posed by German and other foreign technology.

But this was still in the future. In 1895 he was awarded the Merton Postmastership in Science but, as he failed Latin Prose in Responsions, he had to return to Aberystwyth for a further term before retaking them. He passed at a second attempt, but his personal assessment of his performance, 'In my own judgement I made more errors in the second than in the first attempt',[32] reveals the arrogance which many of his acquaintances found repellent, even when they recognised the shyness which often prompted it.

Soddy was now able to begin his undergraduate career at Merton. Even so, there were problems. He arrived a term later than his contemporaries and by the time he appeared all the college rooms were taken. As a result he had to board in the town for the remainder of the first year. He found getting up early to keep roll calls in college at eight o'clock in the morning a cause for resentment, but perhaps the effect on his social life gave the greatest concern. 'By arriving at Merton in January 1896, I found myself the odd man out—a Freshman to my own year and senior to the next'.[33]

Perhaps as a result of this delayed arrival, perhaps because of his shyness, Soddy made few new friends during his time at Oxford. There was, however, an exception. His friend from Eastbourne College, Harold Cort Carpenter, was in his final year at Merton and made a particular effort to befriend the younger boy. Soddy gave the following example of Carpenter's behaviour: 'Harold was always the English gentleman with a universal consideration for others and with complete unselfishness. I was disturbed by his utter defiance of custom when, at hall dinner, in spite of the rule of segregating by years, he always, if I were dining, left his place at the top of the table to sit with me near the bottom'.[34] Carpenter not only sat with Soddy at mealtimes, but also made certain that he had friends with whom to spend his evenings and other spare time. They had more than memories of school days in common as they were also both chemists. The friendship developed during Soddy's first year and was to endure after their Oxford days.

Soddy and Carpenter spent Soddy's two long vacations at Oxford together. In 1896, at the end of Soddy's first year, they both went to Germany. As the recognised home of chemical research, a number of important scientific papers appeared first in German. It was therefore necessary for research scientists to have at least some knowledge of the language to enable them to keep up with current research. As a result of this, a number of British graduate students completed their education there. Carpenter was one of them and after graduating he moved to Germany to learn the language and then to study for his doctorate. During the summer vacation Soddy visited him there and gained a very useful, if limited, understanding of German.[35]

The second long vacation was very different in place and purpose. It could also be relied upon to prepare students for a scientific education and so it was decided that Frederick should go with him for the year and finish his preparations for Oxford there.[28]

The Soddy brothers' time at Aberystwyth was both useful and thoroughly enjoyable. They had rooms on the sea front, ideal for their hobbies of swimming and, for Tom, sea rowing. Cycling through the Welsh countryside was another pleasure for Frederick, who was developing his enduring interest in his surroundings. He also discovered an unexpected and very welcome aspect of the Welsh college. This was attendance at the 'conversaziones' held by the more senior women students. At these events, new students could meet others, friendships could be pursued and much of the loneliness of student life avoided. There was the added advantage for shy young people that the college admitted women as well as men so that young people from their single-sex schools could become acquainted with the other sex in relaxed and comfortable surroundings. Compared with either public school or an Oxford college this was very pleasant and so it is not surprising that Soddy was to recall his days at Aberystwyth as one of the most enjoyable parts of his education.[29] However, when he moved on, the contrast between the friendly and welcoming atmosphere of Aberystwyth and his introduction to Oxford was to be stark indeed.

Life at Aberystwyth was not just social. Frederick was also required to undertake a wide programme of study which included reading modern English literature and the classics in order to improve his style of writing. As this extract from an early example of his work, his submission for the Evans Prize Essay at Aberystwyth in 1895, shows, he needed some help, especially with punctuation,[30]

Although much has been discovered during late years, it seems highly probable that only the boundaries of the subject have been skirted and that the vast field of research on the borderland between chemistry and physics is almost virgin soil, holding out a bountiful harvest to those who, not content with treading the well-beaten paths of science, are enterprising enough to attack the problem, and patient enough to overcome the difficulties, which pioneers of scientific research always have to encounter.

This youthful essay was to prove prophetic. Within ten years of writing it, Soddy was to be 'harvesting' the rewards of his collaboration with Rutherford which was on 'the borderland between physics and chemistry'. By then, his writing had also improved and his clear use of English was given as one explanation for his popularity amongst writers on radioactivity. By the mid-1920s his works had been singled out by the scientific establishment of the USSR for translation because of his 'straightforward style'.[31]

in life with a secure future. The future for a scientist even in the late nineteenth century was far from secure. Career opportunities were few and far between. The Devonshire Commission summed up the position in the 1870s:[24]

It is acknowledged that Science is neither recognised, nor paid nor rewarded, by the State as it ought to be, that mainly owing to this, there is no career for Science and that parents and masters are justified in avoiding it.

This position was changing by the 1890s. In response to the increasing pressure from abroad, especially from Germany, during the second half of the nineteenth century, some attempts had been made to improve scientific education through the growth of Mechanics Institutions, the development of evening classes of various kinds and the founding of various colleges outside the older universities. Examples of these were the City and Guilds College which Frederick had hoped to attend, and the Normal School of Science in South Kensington where H. G. Wells had studied.[25] However, the opportunities for a career after this were still distinctly limited, and even more so after taking a degree in pure rather than applied science. There was also a residual class-based prejudice through the late nineteenth and early twentieth centuries described by Sir Eric Ashby: 'it was admitted that the study of science for its useful application might be appropriate for the labouring classes, but managers were not attracted to the study of science except as an agreeable occupation for their leisure'.[26]

Benjamin Soddy's uncertainty when he was asked to agree to the prospect of a science degree at university for his son was therefore quite justified. While in theory he was prepared to support Frederick as an undergraduate, he quite reasonably wanted to know how soon afterwards he might be self-sufficient. Somewhat optimistically, Hughes assured him that within a couple of years of leaving the university, Frederick would be earning four hundred pounds a year. As a result, Benjamin Soddy agreed to the proposal and the only remaining problem was for Frederick to pass the necessary exams.

Very soon after this, Hughes was appointed HM Inspector of Schools and so would leave Eastbourne College. This raised potential problems since it was unlikely that his replacement would be so particularly suited to Frederick's needs or as able or sympathetic a teacher. Various attempts were made by school and family to find an alternative. The first possible solution was the suggestion that the Oxford scholarship examination should be taken a year early. This was vetoed by the Merton College authorities who felt that at sixteen Frederick was too young.[27] However, his brother, Tom, was going to Aberystwyth to study for his BA as a first step towards eventual ordination. As this was Hughes's old college, it

Frederick Soddy in 1952

The World Made New

Frederick Soddy, Science, Politics, and Environment